CRYPTODIGITS
VOLUME I

KLOOTO Games
San Antonio, Texas

Published by KLOOTO Games
www.orbiculuspublishing.com
Direct all inquiries to:
author@orbiculuspublishing.com
Copyright 2015 Cyrus F. Rea
All rights reserved.
ISBN-13: 978-1516819737
ISBN-10: 151681973X

WELCOME!

Inside, you'll find 200 addition and multiplication problems...the type of problems a third-grader could solve. But, each Digit in each problem has been replaced with a Letter. Your task is to **DECODE** the problem. Each Letter equals one and only one Digit. And, each problem has only one solution.

"But, I hate math!" Oh, don't be a baby. These puzzles aren't about math...They're about **CRYPTOLOGY**. If you know that 2 + 2 = 4, you are more than capable of solving these puzzles.

You'll find four categories: **EASY**, **MEDIUM**, **DIFFICULT**, and **GENIUS LEVEL**. The first two categories give you the Digits used as well as a handy grid to help in your deductions. As far as the last two categories go, you are on your own...some of them are deviously horrendous.

But, KLOOTO has provided **HINTS**! There are two types of hints given in the back of this book. Feel free to use them. Even with the hints, you'll have plenty of code-cracking left to do. The **SOLUTIONS** follow the **HINTS**.

Finally, if you have absolutely no idea how to get started, KLOOTO has provided a complete **SAMPLE PROBLEM** immediately following this introduction. This walk-through will get you up to speed.

So, sit back, grab a pencil, and allow your brain to do its magic!

SAMPLE PROBLEM

Let's take a tour of a problem.

On the left is the actual problem that needs decoding.

In the middle -- at least on the easier problems -- you'll find a handy grid to help you eliminate possibilities. This grid also tells you what digits are used.

Lastly, KLOOTO has included some boxes where you can fill in your answers and guesses.

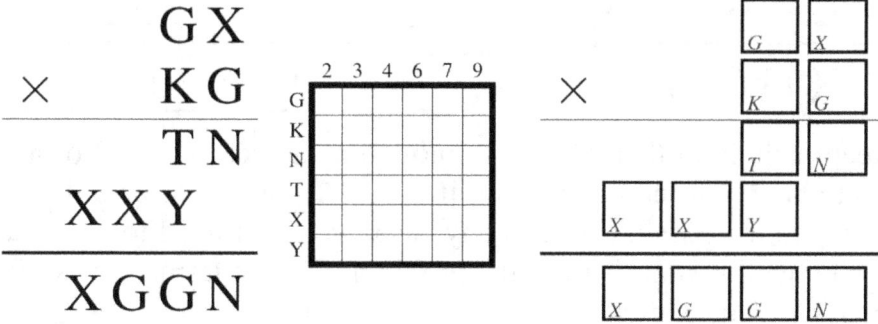

Where to start?! Well, each problem is different, and there are numerous ways to attack each problem. Outright guessing *usually* isn't necessary... just logic and deduction. For instance, take a look at the addition portion of the problem. Specifically, it appears that **T** plus **Y** results in **G**. But, there also must have been a **carry** because the second **X** in **XXY** turns into a **G**. Now, the **carry** number can only be a **1**. (See why? The largest number **T** and **Y** could ever add up to is "16".) So, if **X** was **9**, then that second **X** would add with the carry and result in **10**. That is, the **G** would be a **0**, and then another **1** would carry over to the left-most **X**. But, we know **G** can't be **0** because there are no zeroes in this problem, and we know nothing carried over to the left-most **X** because it remains unchanged when it's all added up. So, let's mark an *X* on our grid to remind ourselves that **X** cannot be **9**.

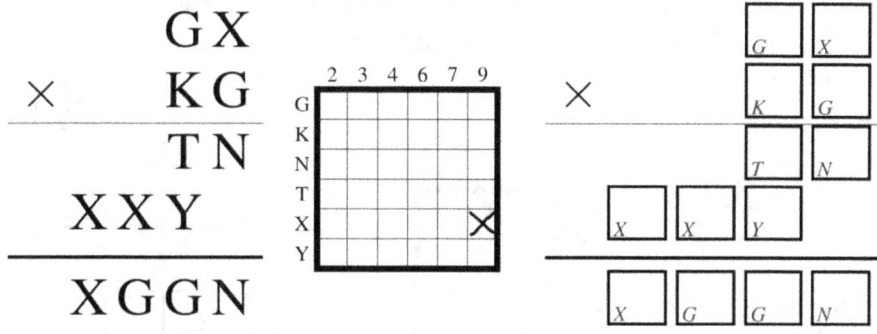

Let's take a look at the **G**. Up at the top, you'll notice that **GX** times **G** equals **TN**. Well, **TN** only has two digits. So, **G** can't be, for instance, **7**. See why? If it was 7 then **GX** times **G** would result in at least 490 something... but, that takes three digits. Using the same logic, we can quickly deduce that **G** can't be **4, 6, 7,** or **9.** Let's mark those off...

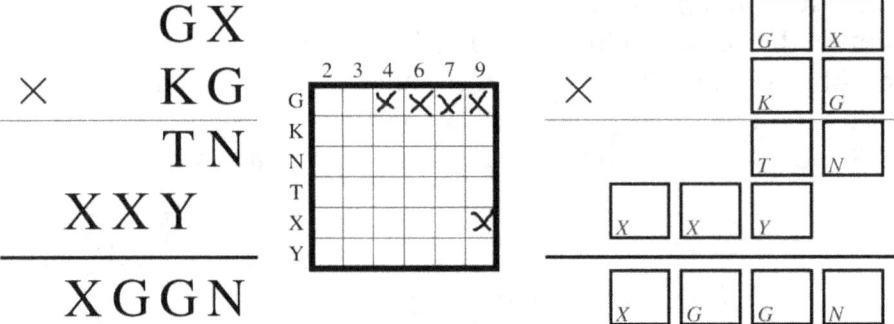

Let's keep picking at that **TN**... We know that **G** has to be a **2** or a **3**. If **G** is a **3**, then the **T** has to be **9**. On the other hand, if **G** is a **2**, the **T** can be a **4** or, *maybe*, a **5** if that **X** is big enough to cause a carry. In any case, we don't need to worry about **T** being a **5** because we know that digit isn't in the problem. So, to summarize, we know that **T** is either a **4** or a **9**. Let's mark it up...

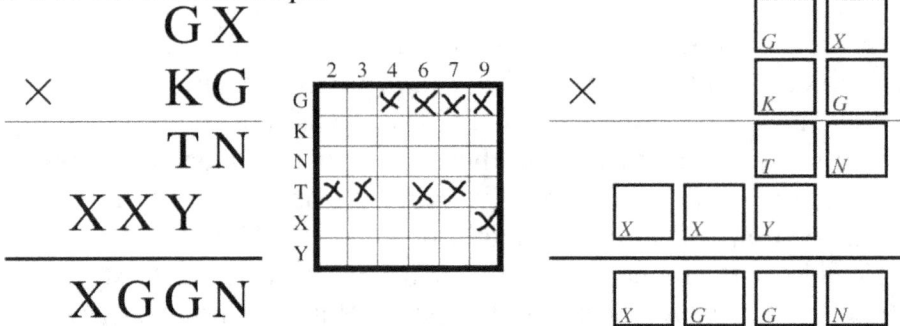

Hmm... Where to next? Well, going back to that **XXY** on the bottom, we know that the second **X** turned into a **G** because of a carry. That means **G** must be one more than **X**. Or, to state it differently, **X** must be one less than **G**. Well, look at our grid! The only way we can have an **X** that is one less than **G** is if **G** is **3** and **X** is **2**! Let's fill it in! (We put circles for matches.... remember when you place a circle, you can fill its row and column in with *X's*).

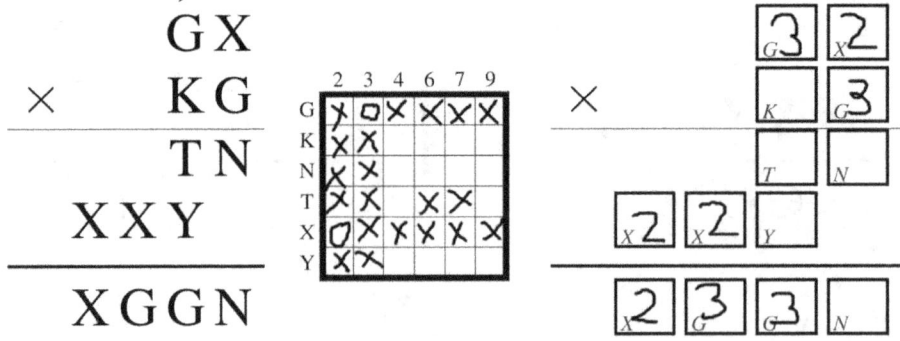

Now we are getting somewhere! We know that **TN** is **96** because we know that **GX** is **32** and **G** is **3**! Let's fill everything in:

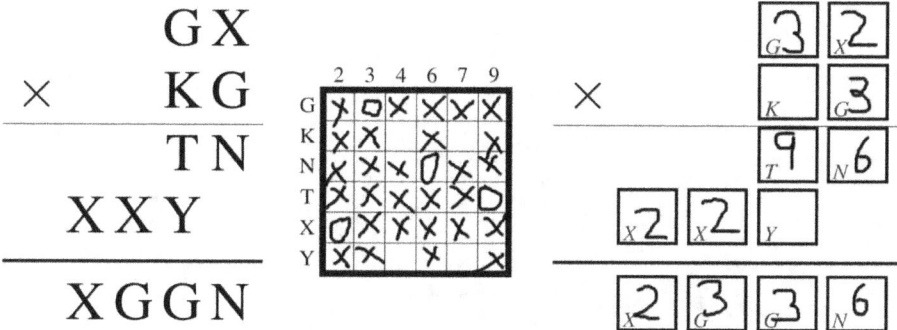

As for the rest? Even KLOOTO can figure it out... **Y** must be **4** and that means that **K** has to be **7**. *Voila*! We did it!!

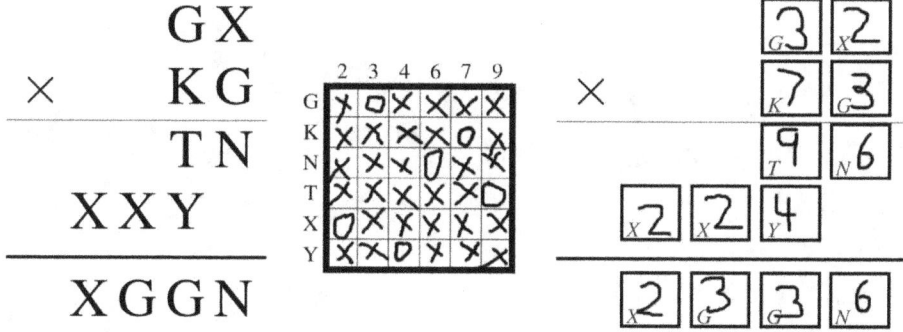

Now, remember... There are a gazillion ways to approach these problems. You may find some 'techniques' more natural than others. But, rest assured that there will be one and only one solution to each problem.

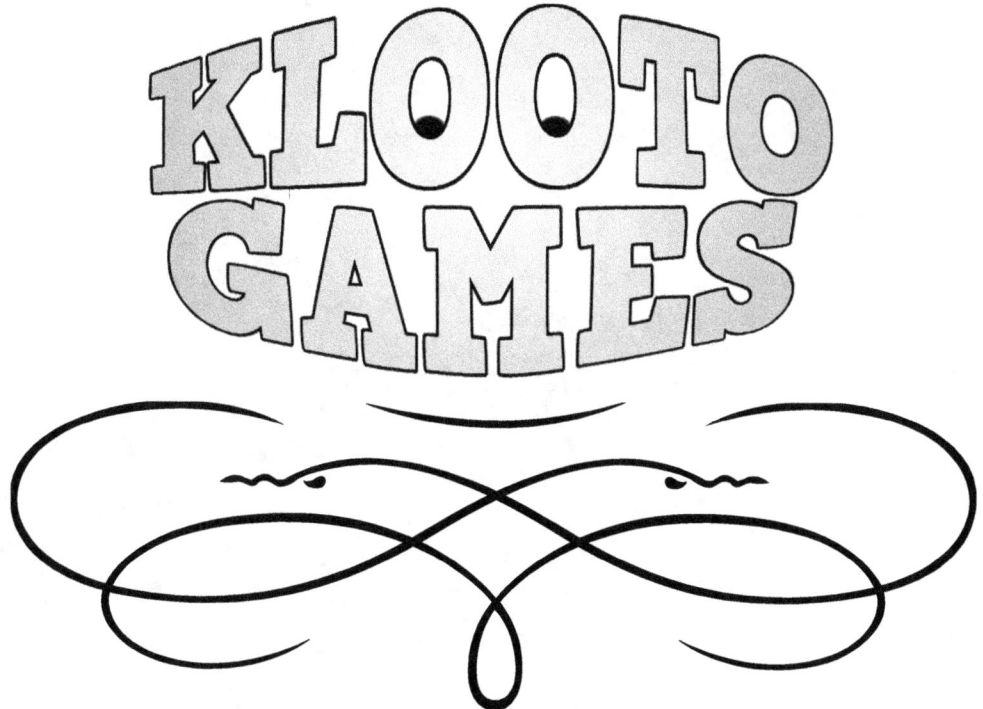

EASY

Problem No. 1:

$$
\begin{array}{r}
ATA \\
\times \quad EG \\
\hline
BEG \\
HTE \quad \\
\hline
GTTG
\end{array}
$$

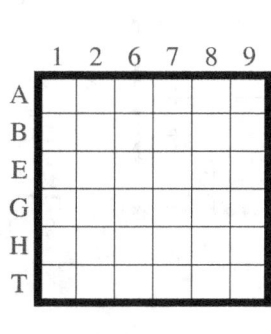

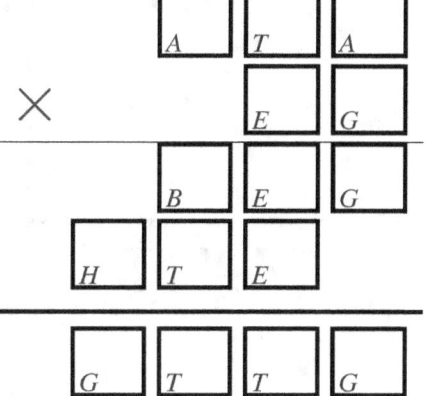

	1	2	6	7	8	9
A						
B						
E						
G						
H						
T						

Problem No. 2:

$$
\begin{array}{r}
NTA \\
+ \quad BAN \\
\hline
TTGT
\end{array}
$$

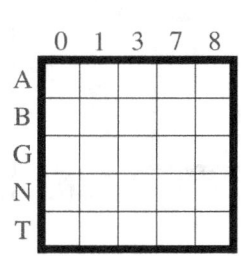

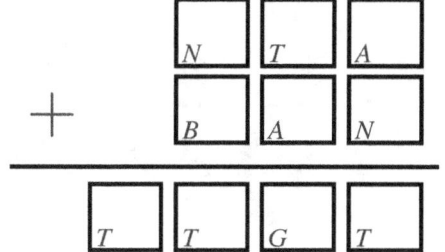

	0	1	3	7	8
A					
B					
G					
N					
T					

Problem No. 3:

$$
\begin{array}{r}
XGG \\
\times \quad NX \\
\hline
GAA \\
XKNB \quad \\
\hline
XXGGA
\end{array}
$$

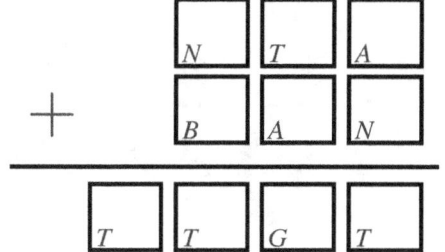

	1	2	4	6	8	9
A						
B						
G						
K						
N						
X						

Problem No. 4:

$$
\begin{array}{r}
G\,B \\
\times\quad B\,G \\
\hline
K\,T \\
T\,X \\
\hline
E\,B\,T
\end{array}
$$

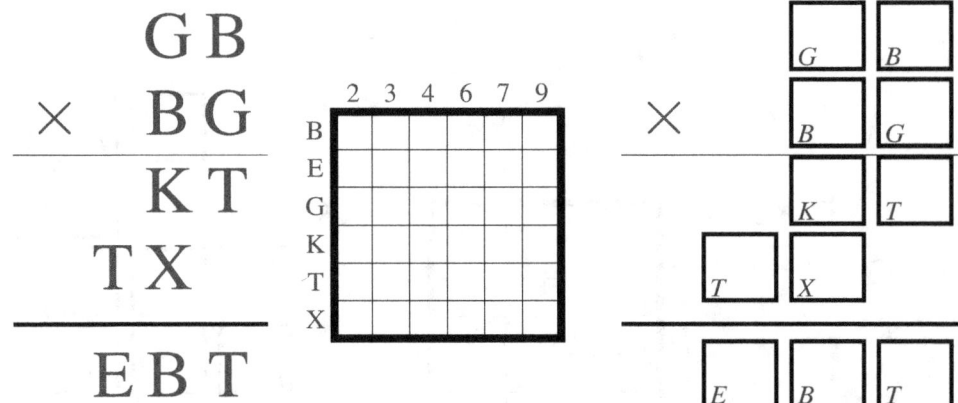

Problem No. 5:

$$
\begin{array}{r}
T\,K\,K \\
+\,T\,K\,T \\
\hline
X\,A\,X
\end{array}
$$

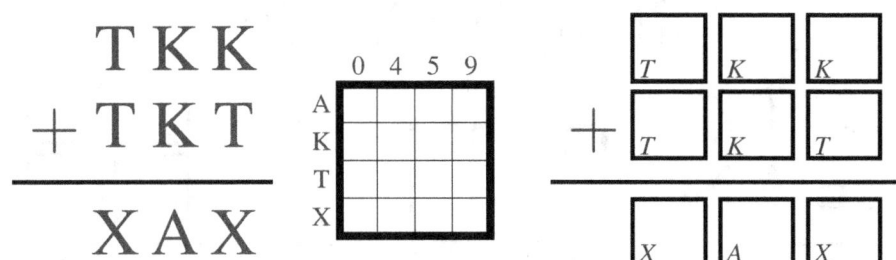

Problem No. 6:

$$
\begin{array}{r}
Y\,H\,H \\
\times\quad A\,N \\
\hline
K\,A\,X\,N \\
K\,X\,H\,X \\
\hline
K\,H\,K\,Y\,N
\end{array}
$$

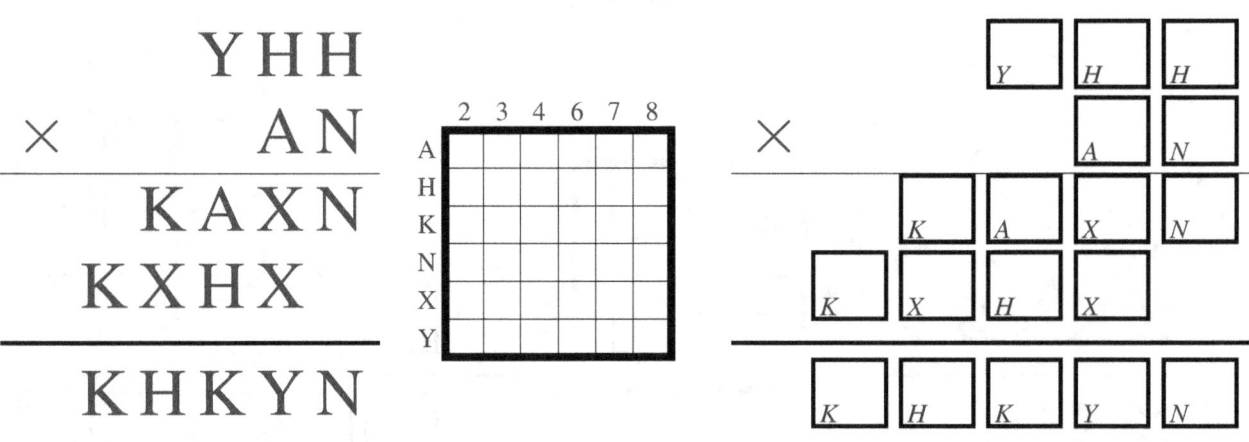

Problem No. 7:

$$
\begin{array}{r}
B\,B\,K \\
+\;\;K\,A\,A \\
\hline
A\,A\,Y\,B
\end{array}
$$

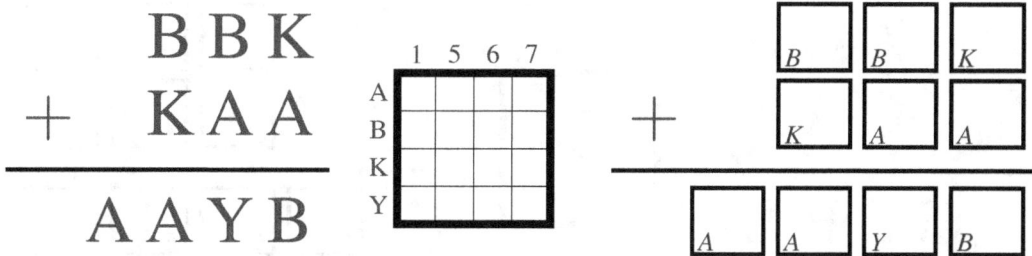

Problem No. 8:

$$
\begin{array}{r}
N\,X\,A \\
\times\;\;\;\;T\,X \\
\hline
T\,G\,A\,N \\
N\,X\,A \\
\hline
E\,N\,N\,N
\end{array}
$$

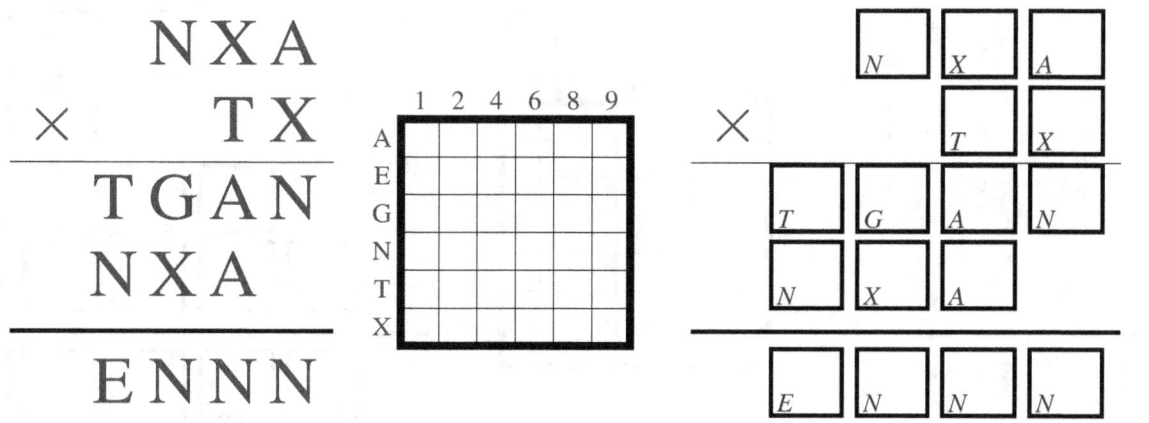

Problem No. 9:

$$
\begin{array}{r}
T\,N\,T \\
\times\;\;\;\;X\,G \\
\hline
N\,X\,H \\
K\,K\,T\,N \\
\hline
K\,T\,K\,T\,H
\end{array}
$$

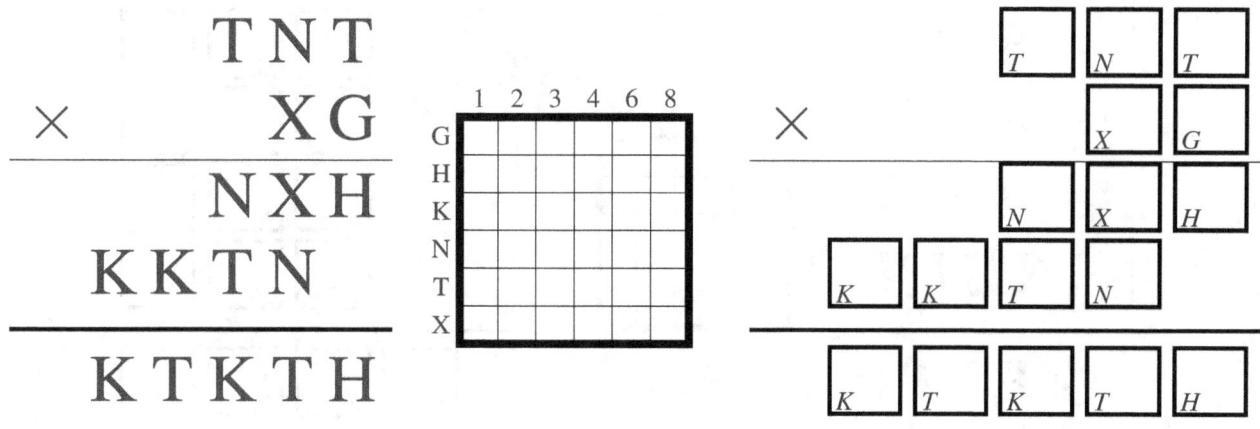

Problem No. 10:

$$\begin{array}{r} A\,G\,B \\ +\ G\,A\,A \\ \hline G\,E\,E\,E \end{array}$$

	0	1	2	8
A				
B				
E				
G				

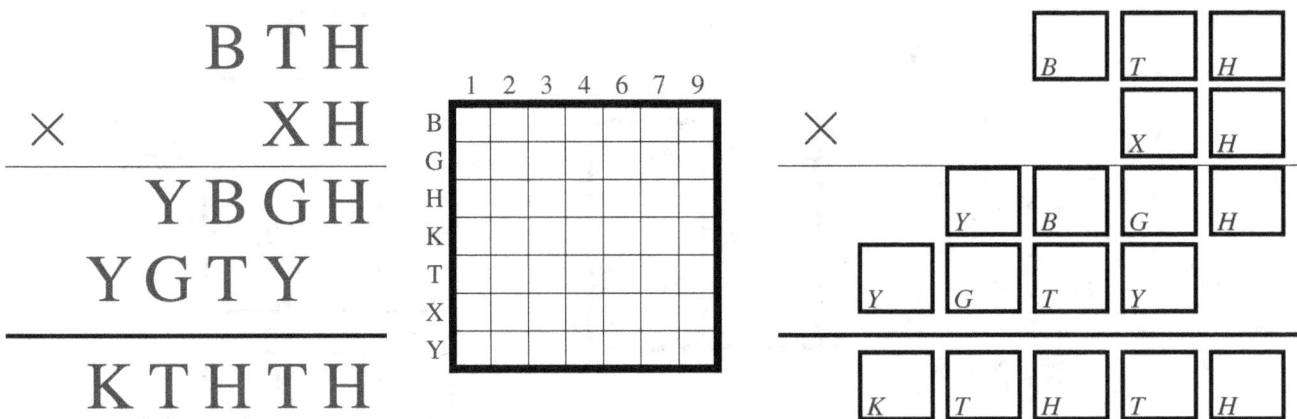

Problem No. 11:

$$\begin{array}{r} B\,T\,H \\ \times\ \ X\,H \\ \hline Y\,B\,G\,H \\ Y\,G\,T\,Y \\ \hline K\,T\,H\,T\,H \end{array}$$

	1	2	3	4	6	7	9
B							
G							
H							
K							
T							
X							
Y							

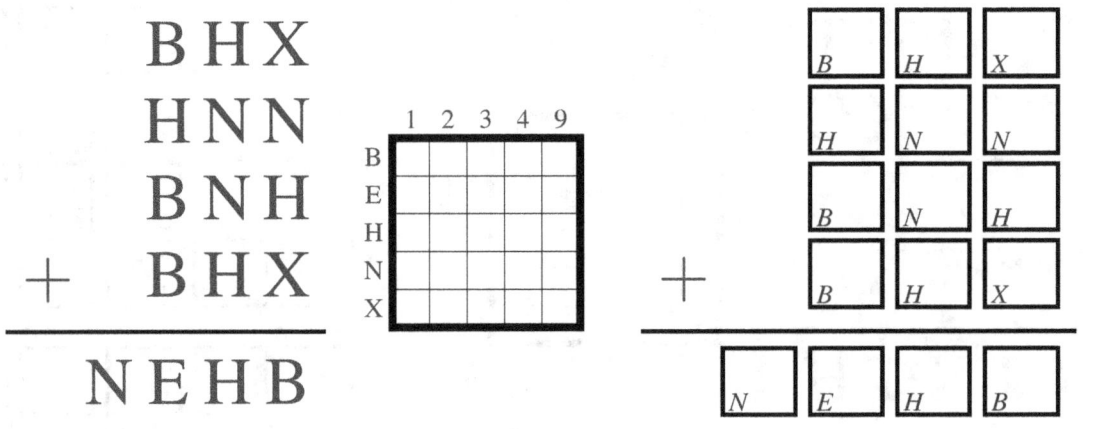

Problem No. 12:

$$\begin{array}{r} B\,H\,X \\ H\,N\,N \\ B\,N\,H \\ +\ B\,H\,X \\ \hline N\,E\,H\,B \end{array}$$

	1	2	3	4	9
B					
E					
H					
N					
X					

Problem No. 13:

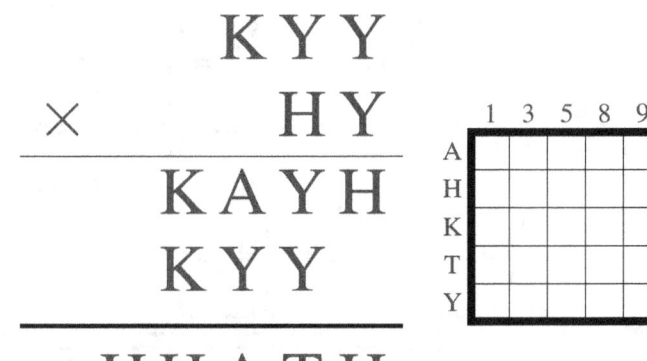

```
    K Y Y
  ×   H Y
  ─────────
  K A Y H
  K Y Y
  ─────────
  H H A T H
```

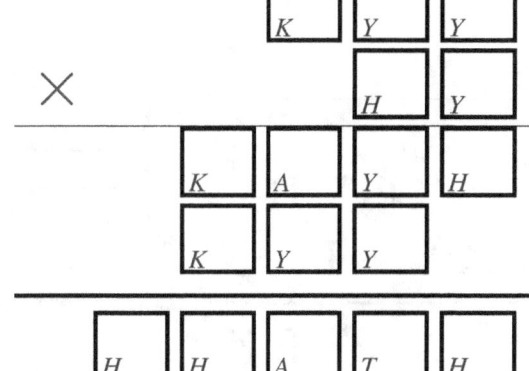

Problem No. 14:

```
  N E A E
  T A A T
  N T Y N
+ X Y Y N
─────────
  E X X Y T
```

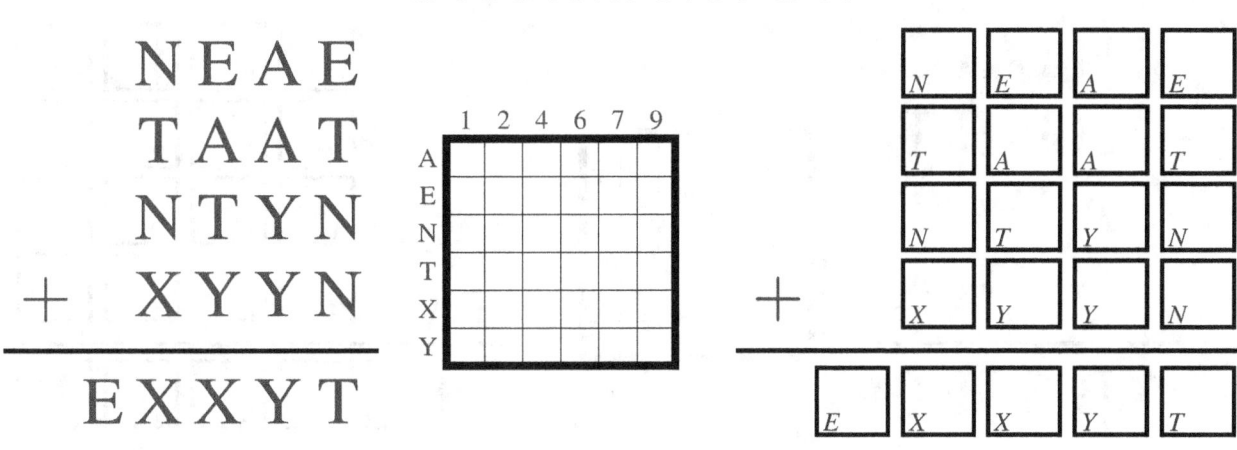

Problem No. 15:

```
  X A N G
+ B Y N Y
─────────
  A G G Y X
```

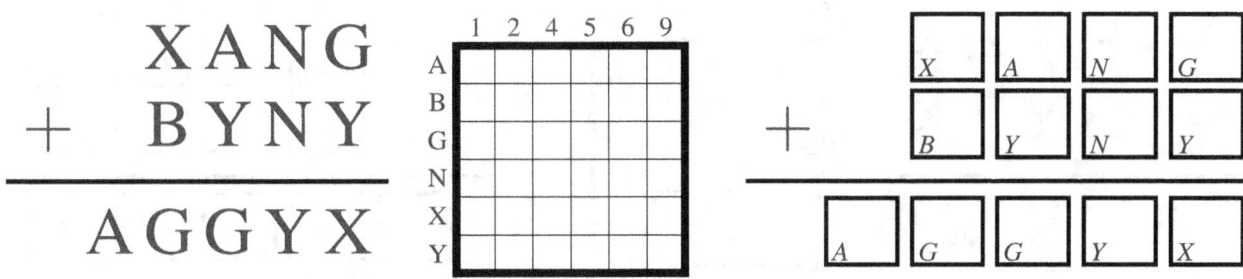

Problem No. 16:

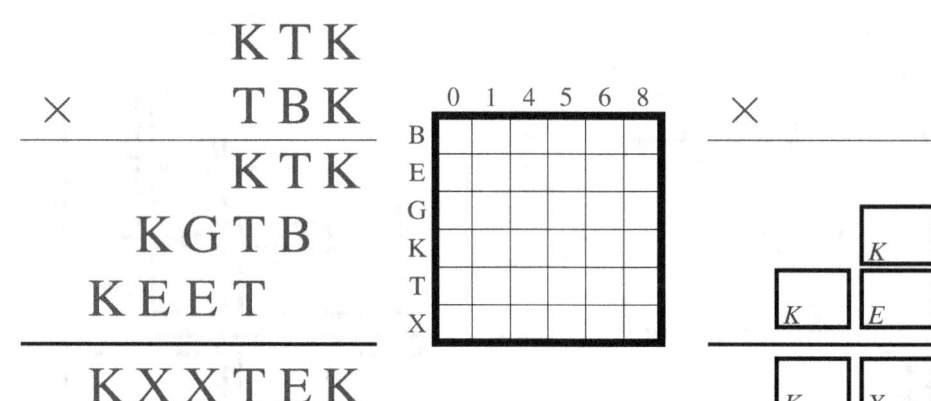

```
        K T K
    ×   T B K
    ─────────
        K T K
      K G T B
    K E E T
    ─────────
    K X X T E K
```

	0	1	4	5	6	8
B						
E						
G						
K						
T						
X						

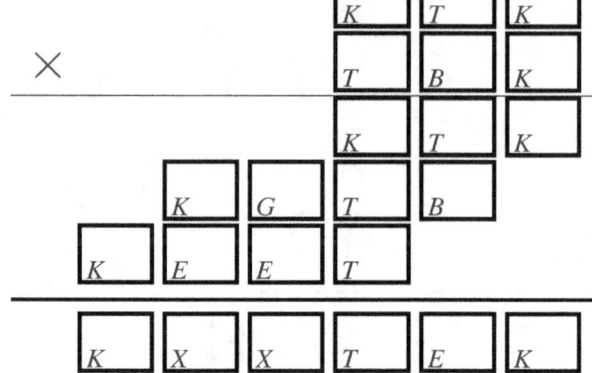

Problem No. 17:

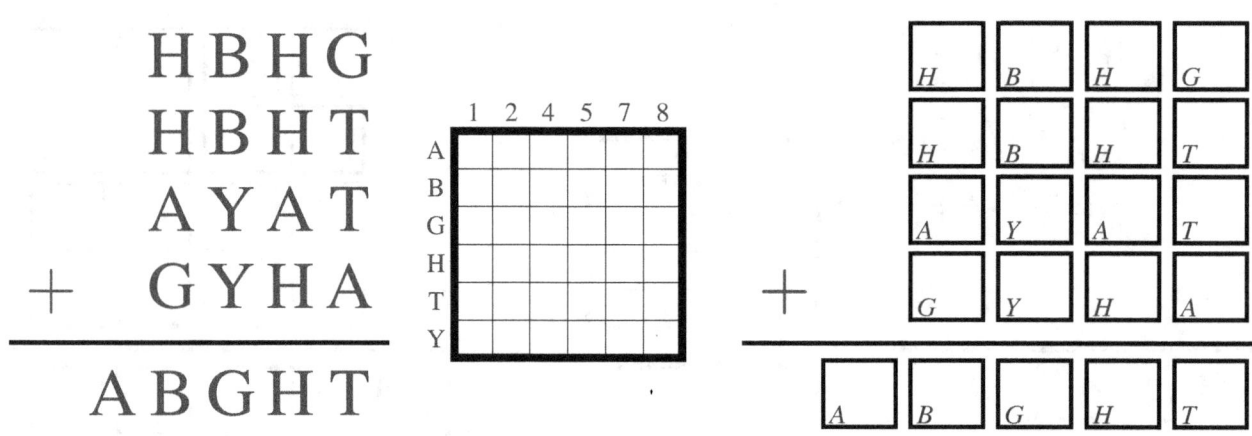

```
    H B H G
    H B H T
    A Y A T
  + G Y H A
  ─────────
    A B G H T
```

	1	2	4	5	7	8
A						
B						
G						
H						
T						
Y						

Problem No. 18:

```
    H T H X
  + H Y Y K
  ─────────
    T G K X H
```

	1	3	4	6	8	9
G						
H						
K						
T						
X						
Y						

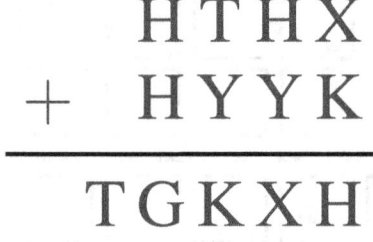
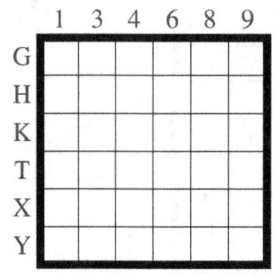
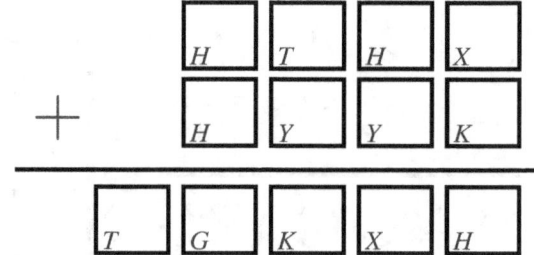

Problem No. 19:

```
  K Y G B
+ A Y A A
─────────
G A H K Y
```

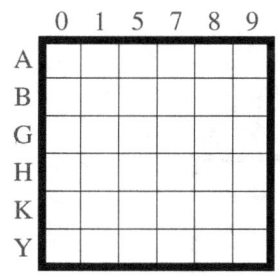

Problem No. 20:

```
  T Y X Y
  A N X T
  X B T A
+ Y B A A
─────────
  N A A N T
```

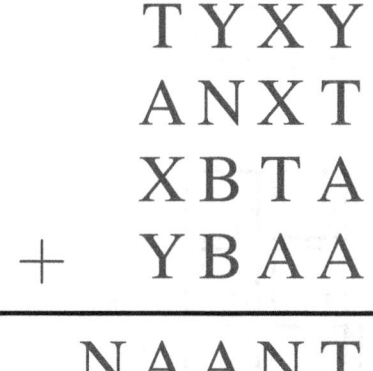

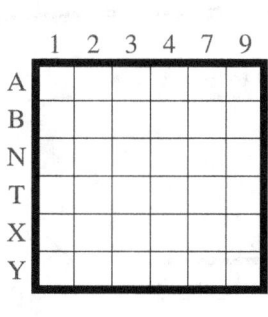

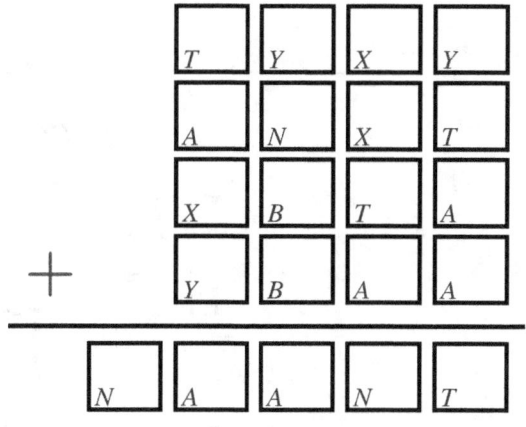

Problem No. 21:

```
    A H
×   Y H
─────────
  A T Y
  A H
─────────
Y Y B Y
```

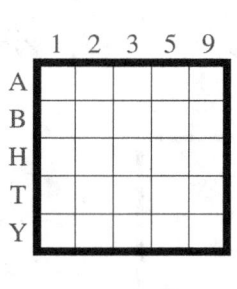

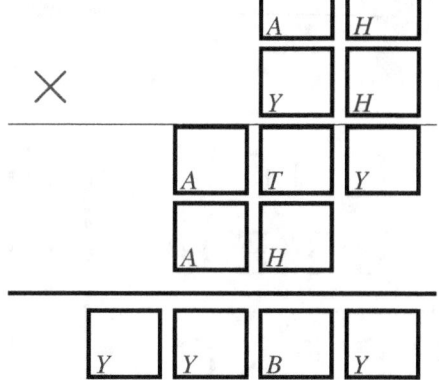

Problem No. 22:

```
    E X X
  ×  T T T
  ─────────
    E X X
  E X X
E X X
─────────────
G E A G X
```

	1	2	5	6	8
A					
E					
G					
T					
X					

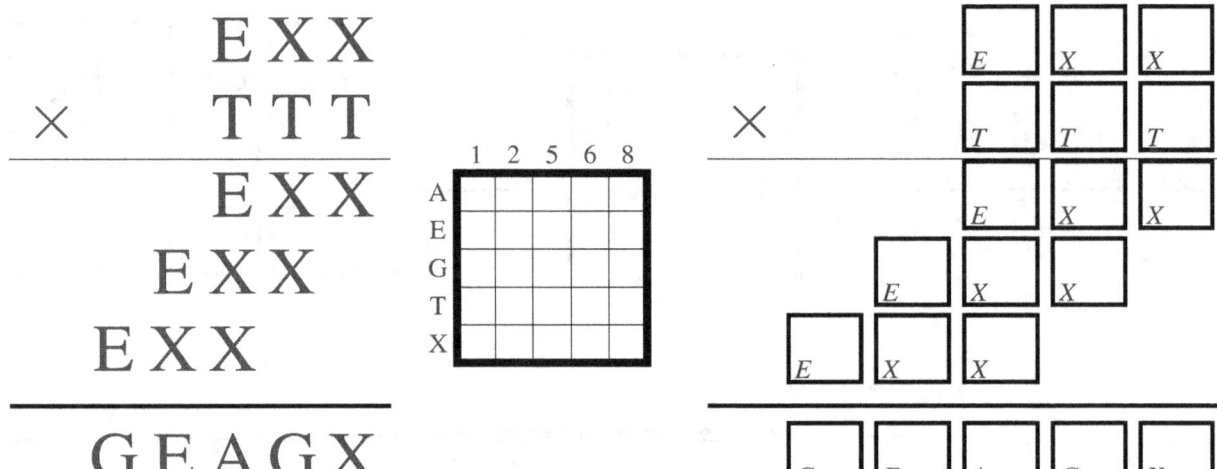

Problem No. 23:

```
    B H
  ×  B H
  ─────────
    B K B
  B H
  ─────────
  T X B
```

	1	3	6	7	9
B					
H					
K					
T					
X					

  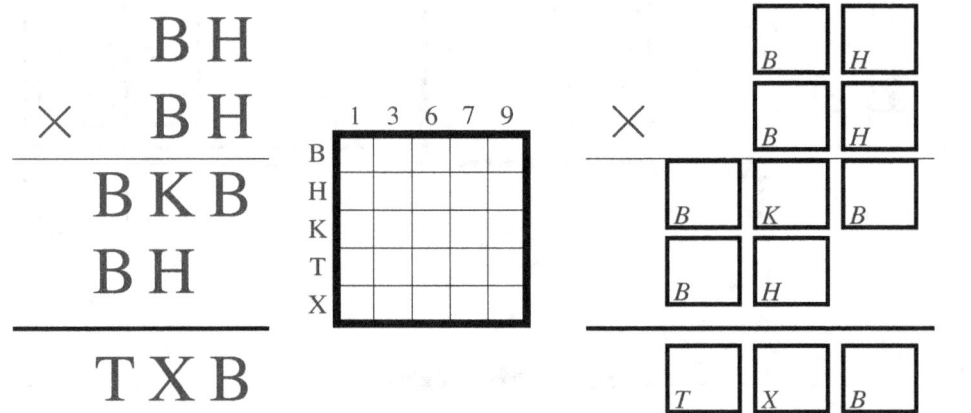

Problem No. 24:

```
    K H G T
  +  B T E E
  ─────────────
  B B E K G
```

	1	3	5	7	8	9
B						
E						
G						
H						
K						
T						

  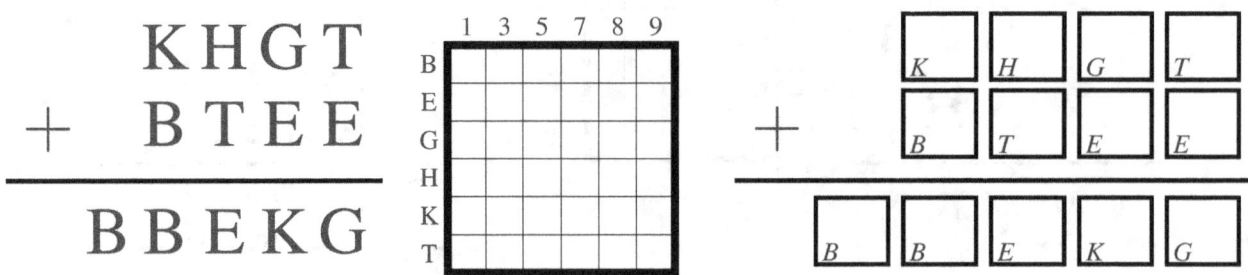

Problem No. 25:

$$
\begin{array}{r}
\text{N T} \\
\times \quad \text{E K} \\
\hline
\text{X N} \\
\text{G Y} \\
\hline
\text{T K N}
\end{array}
$$

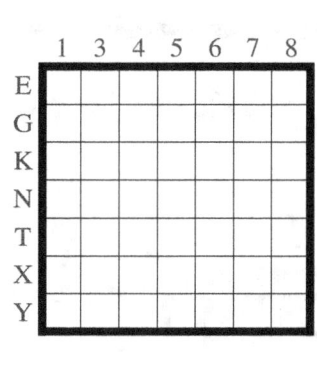

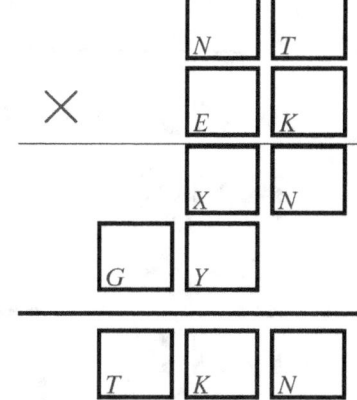

Problem No. 26:

$$
\begin{array}{r}
\text{E Y N} \\
\times \quad \text{Y G Y} \\
\hline
\text{A A K A} \\
\text{E Y N} \\
\text{A A K A} \\
\hline
\text{A E A E E A}
\end{array}
$$

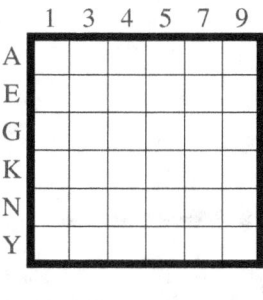

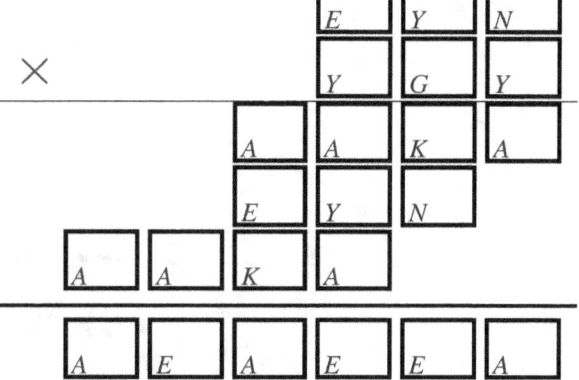

Problem No. 27:

$$
\begin{array}{r}
\text{Y A X G} \\
\text{K Y X X} \\
\text{N G N G} \\
+ \quad \text{Y A Y G} \\
\hline
\text{K G G N A}
\end{array}
$$

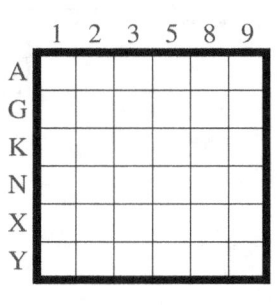

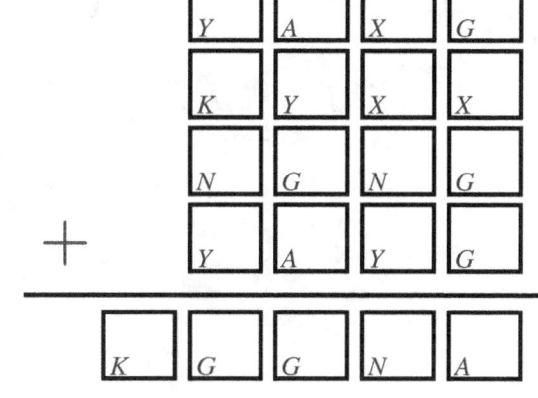

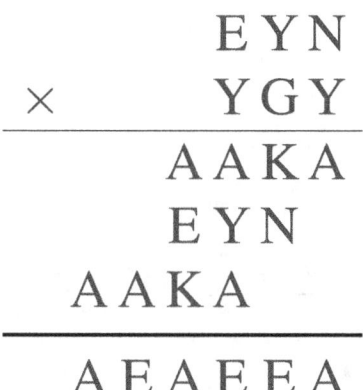

Problem No. 28:

```
    E H N
×   E K N
  ─────────
  E E N H
  H H H
  E H N
  ─────────
  X B H X H
```

	0	1	2	3	4	8
B						
E						
H						
K						
N						
X						

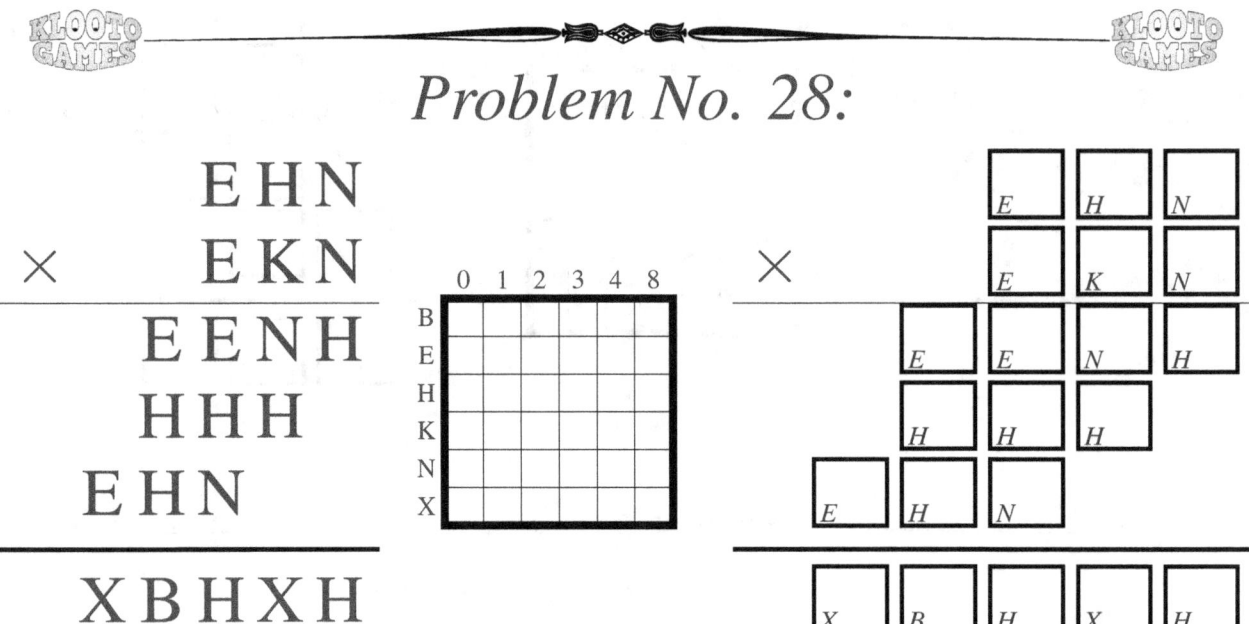

Problem No. 29:

```
    A Y N
×     Y A
  ─────────
  B G Y A
  K B B Y
  ─────────
  K Y K A A
```

	0	1	4	5	6	8
A						
B						
G						
K						
N						
Y						

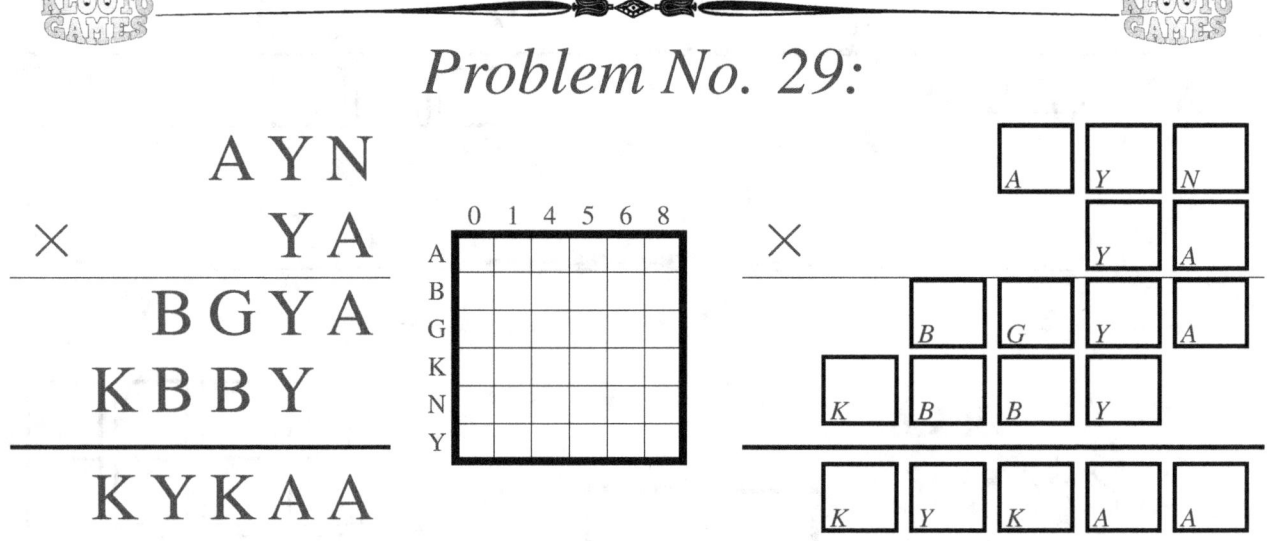

Problem No. 30:

```
        H H N
   ×    K K Y
  ───────────
        H H N
      E H K N
    E H K N
  ───────────
    E B Y N H N
```

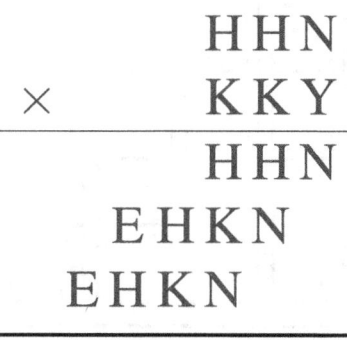
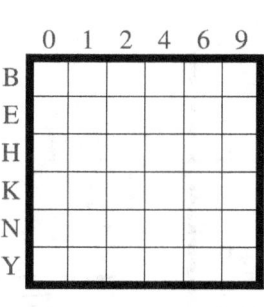
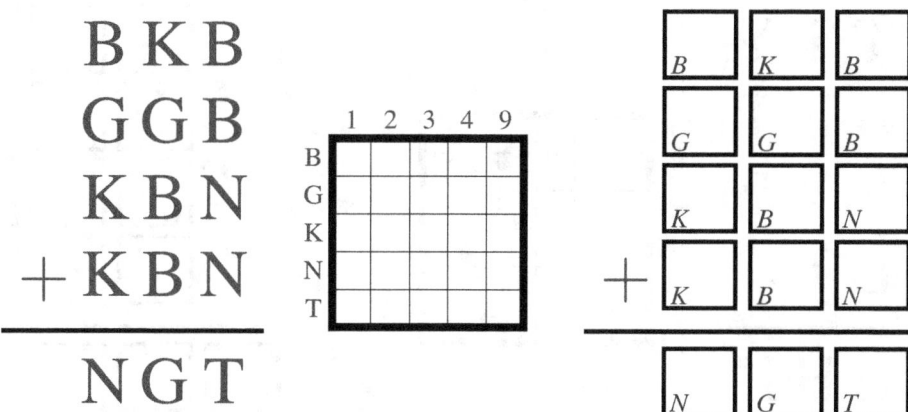

Problem No. 31:

```
    B K B
    G G B
    K B N
  + K B N
  ───────
    N G T
```

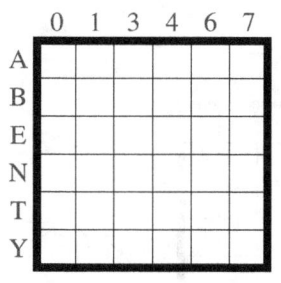
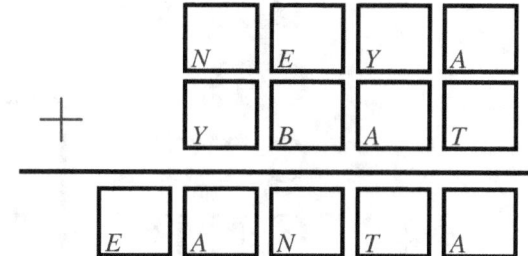

Problem No. 32:

```
    N E Y A
  + Y B A T
  ─────────
    E A N T A
```

Problem No. 33:

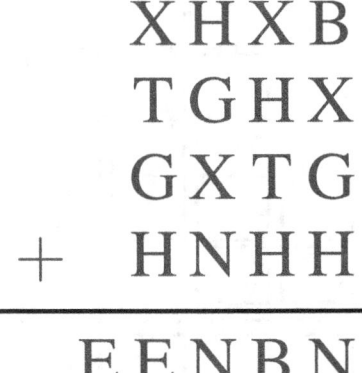

```
  X H X B
  T G H X
  G X T G
+ H N H H
─────────
  E E N B N
```

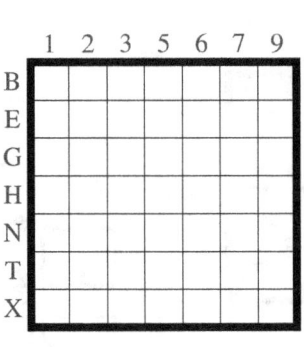

Problem No. 34:

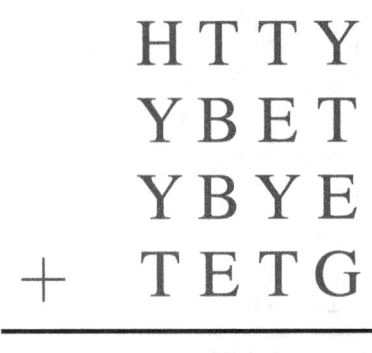

```
  H T T Y
  Y B E T
  Y B Y E
+ T E T G
─────────
  B Y H H H
```

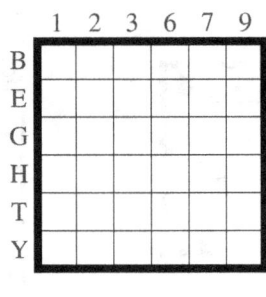

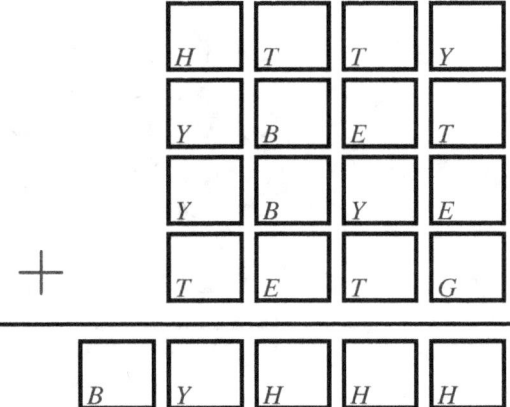

Problem No. 35:

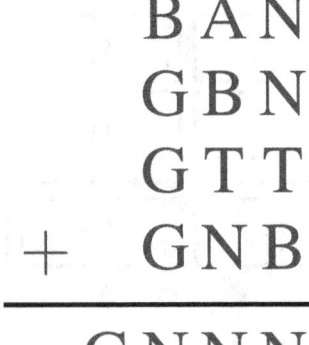

```
  B A N
  G B N
  G T T
+ G N B
─────────
  G N N N
```

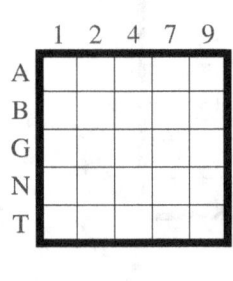

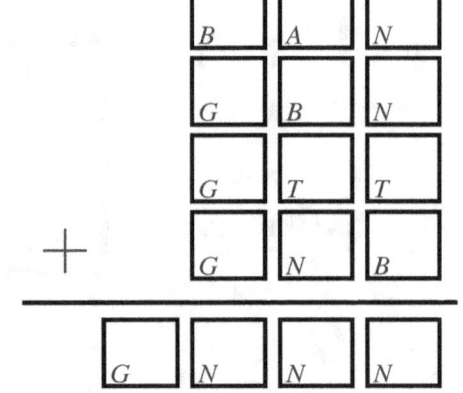

Problem No. 36:

```
  X Y N Y
  B X B T
  N H Y B
+ H H B N
─────────
  X T H B B
```

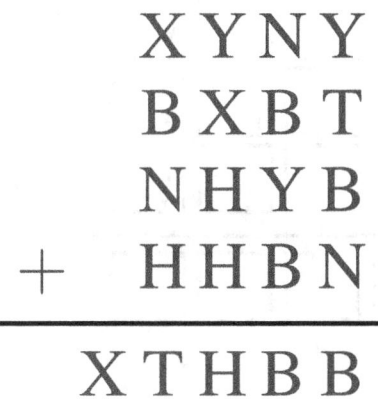

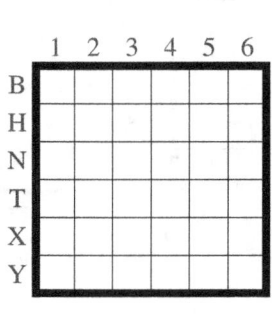

Problem No. 37:

```
  B B N T
+ A N G A
─────────
  T N Y T Y
```

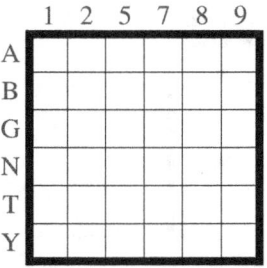

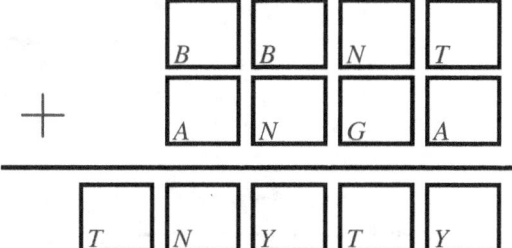

Problem No. 38:

```
      Y G
  ×   H H
  ───────
    A B Y
    A B Y
  ───────
  A Y H Y
```

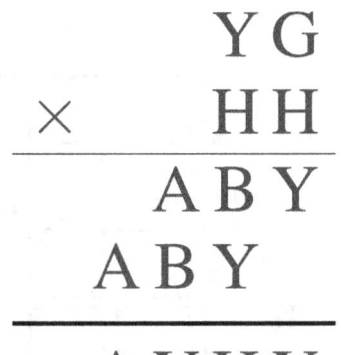

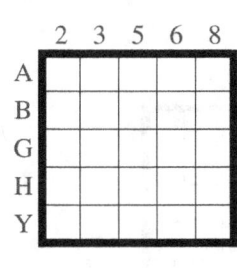

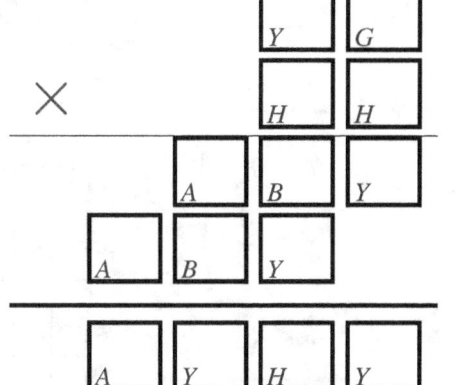

Problem No. 39:

```
    B B E
    Y K E
    K Y E
 +  Y Y N
 ─────────
  N K B B
```

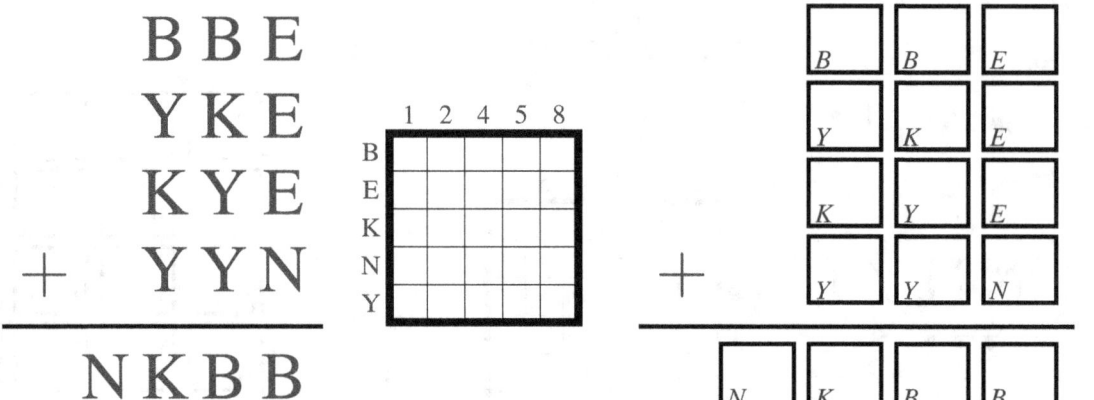

Problem No. 40:

```
    N N A
    G G Y
    Y E E
 +  E G Y
 ─────────
  N A E E
```

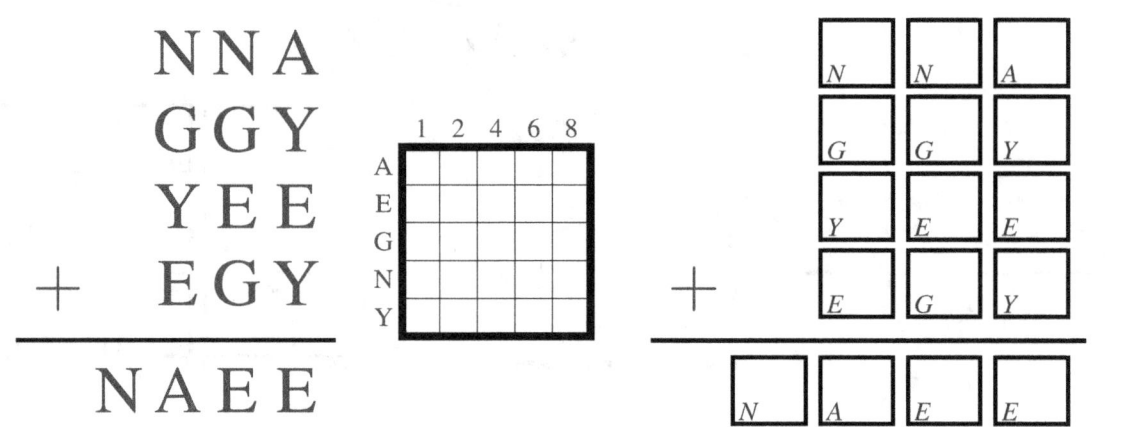

Problem No. 41:

```
    K E E
 ×    X B
 ─────────
    K E E
  B E Y H
 ─────────
  B K H B E
```

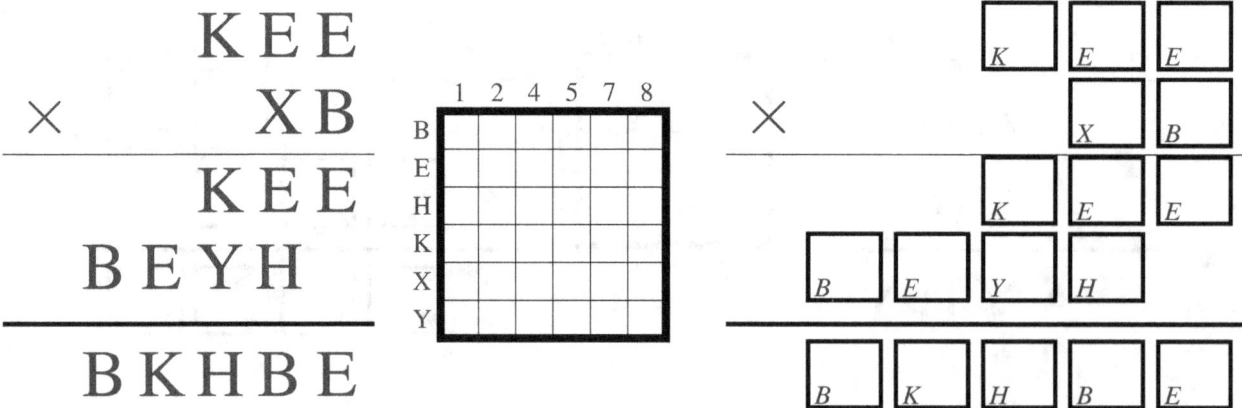

Problem No. 42:

```
    A B A
  ×   X Y
  -------
  A A T E
  H B T
  -------
  Y E A E
```

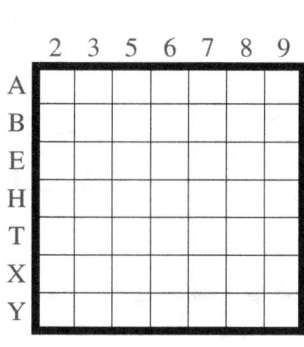

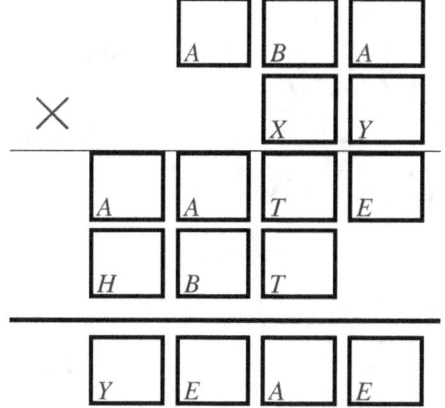

Problem No. 43:

```
    X N X A
    N N N A
    Y T Y N
  + B A N B
  ---------
  T X Y X X
```

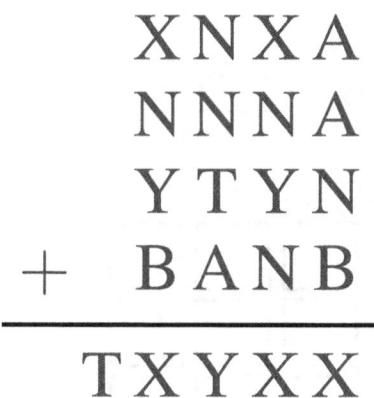

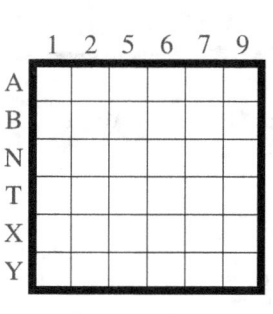

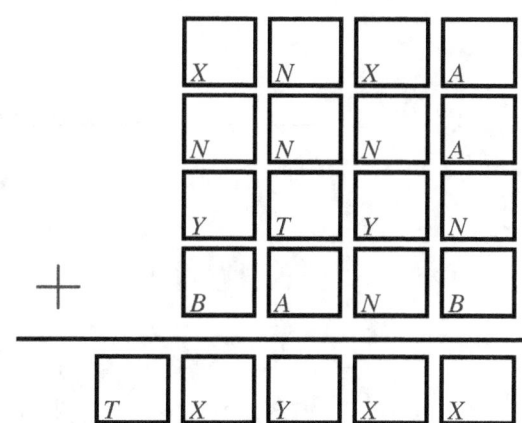

Problem No. 44:

$$
\begin{array}{r}
B\,K\,G \\
\times \quad N\,N\,A \\
\hline
G\,Y\,G \\
B\,G\,K\,G \\
B\,G\,K\,G \\
\hline
B\,K\,A\,K\,K\,G
\end{array}
$$

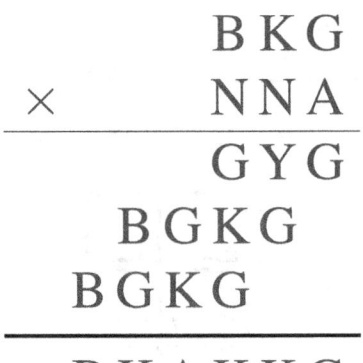

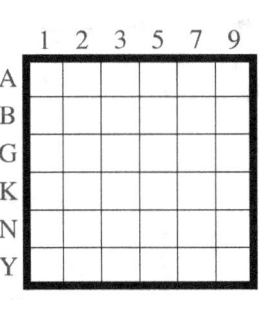

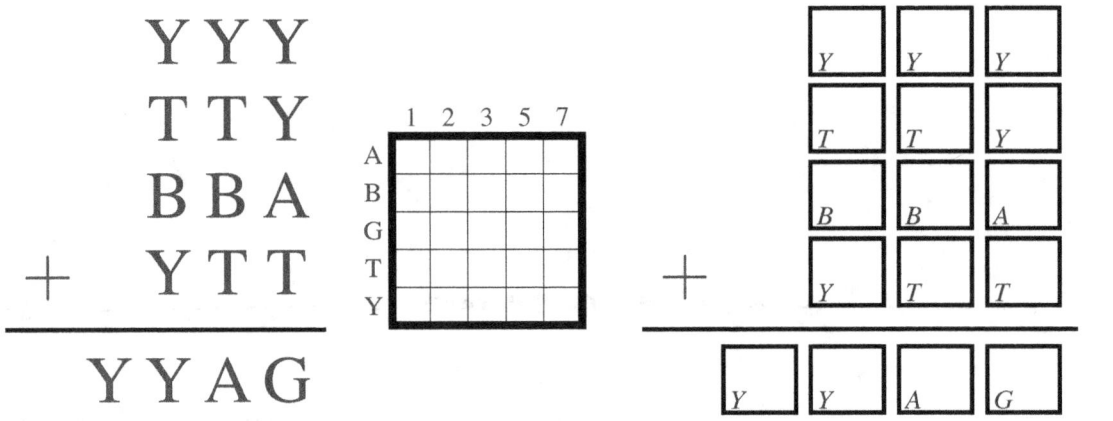

Problem No. 45:

$$
\begin{array}{r}
Y\,Y\,Y \\
T\,T\,Y \\
B\,B\,A \\
+ \quad Y\,T\,T \\
\hline
Y\,Y\,A\,G
\end{array}
$$

Problem No. 46:

$$
\begin{array}{r}
X\,Y\,H \\
X\,X\,X \\
K\,A\,K \\
+ \quad K\,H\,Y \\
\hline
H\,A\,H\,K
\end{array}
$$

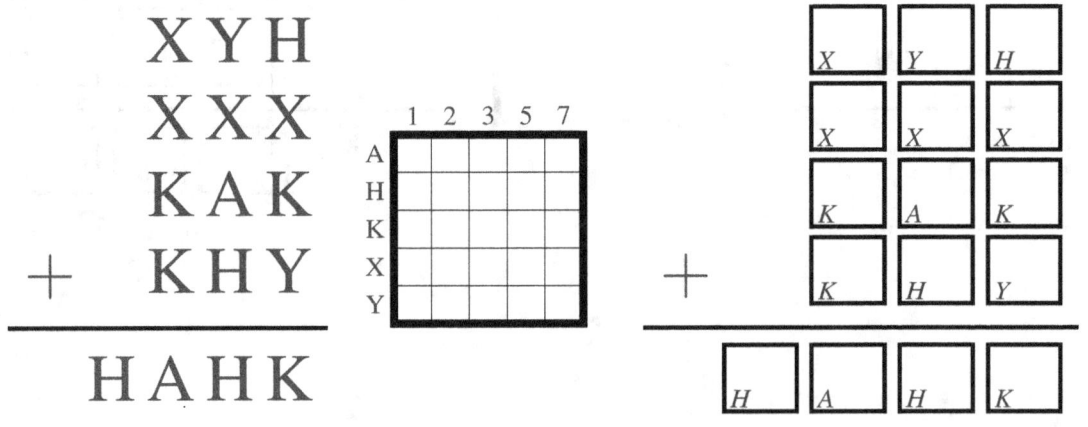

Problem No. 47:

```
    K B N
  ×   K G
  ─────────
    X X B
  K B N
  ─────────
  N B K B
```

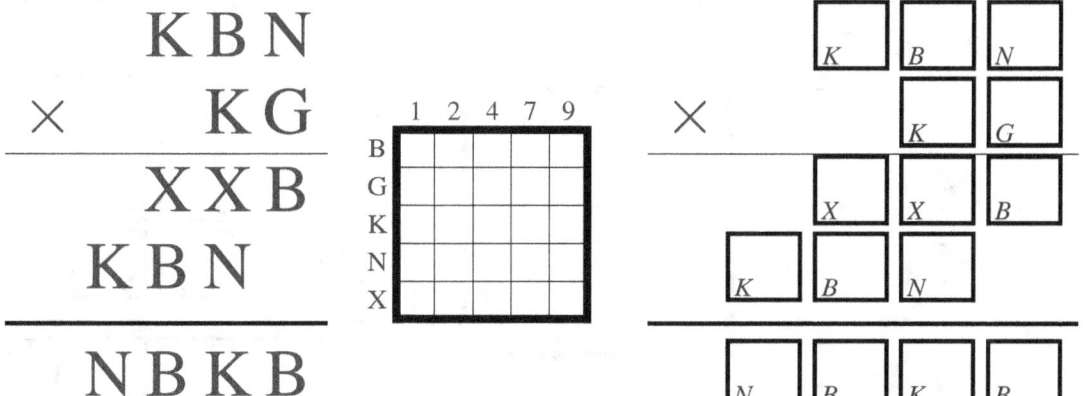

	1	2	4	7	9
B					
G					
K					
N					
X					

Problem No. 48:

```
  B X K
  T X B
  K T X
+ K K B
─────────
  T T T X
```

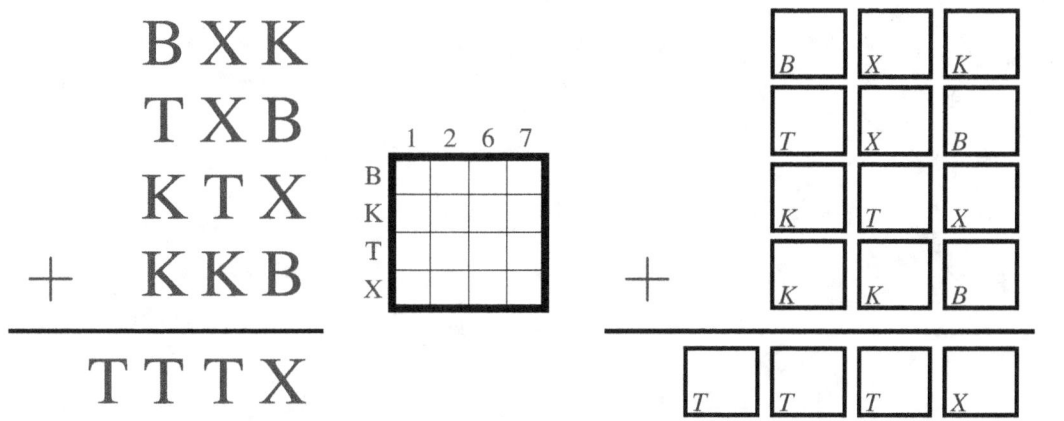

	1	2	6	7
B				
K				
T				
X				

Problem No. 49:

```
  A N Y B
+ K H Y B
─────────
  H A G A N
```

	0	1	2	4	6	8	9
A							
B							
G							
H							
K							
N							
Y							

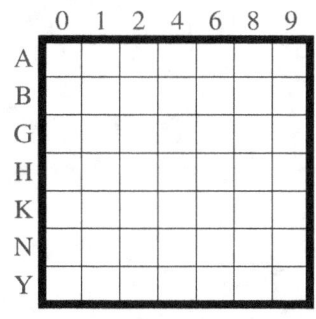

Problem No. 50:

$$
\begin{array}{r}
B\,B\,K \\
\times \quad K\,Y \\
\hline
Y\,E\,G \\
K\,K\,H \\
\hline
K\,N\,Y\,G
\end{array}
$$

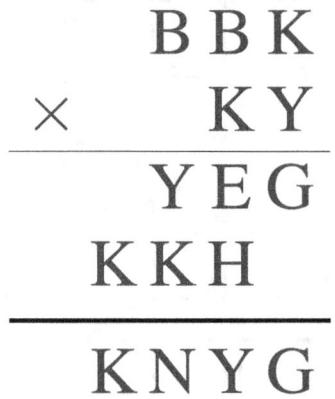

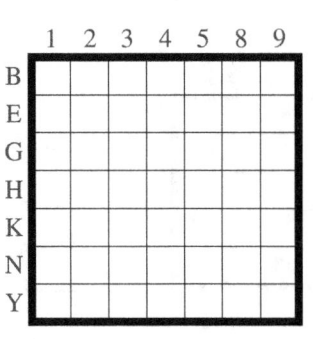

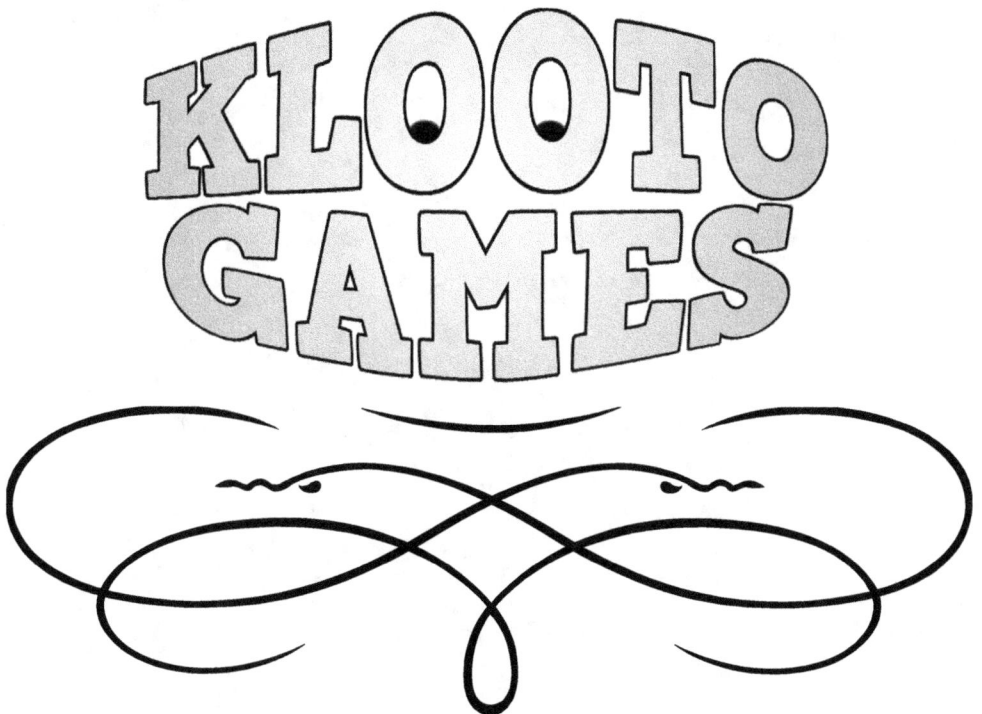

MEDIUM

Problem No. 51:

$$
\begin{array}{r}
B X X H \\
X H K T \\
+ \quad T K K H \\
\hline
N Y H B X
\end{array}
$$

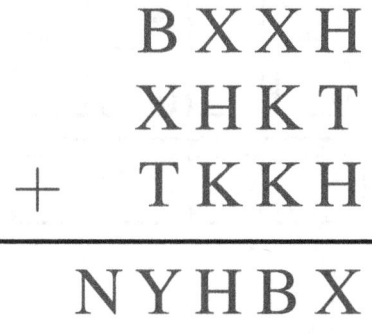

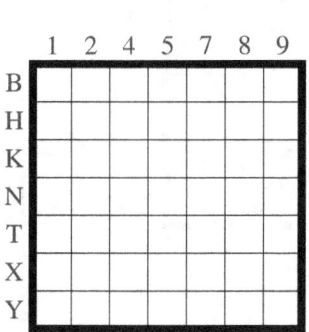

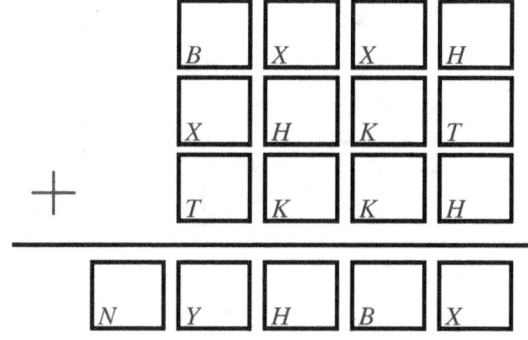

Problem No. 52:

$$
\begin{array}{r}
E T N X \\
K E E E \\
+ \quad N G K T \\
\hline
G K X N E
\end{array}
$$

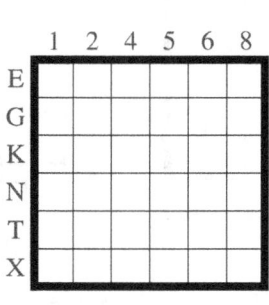

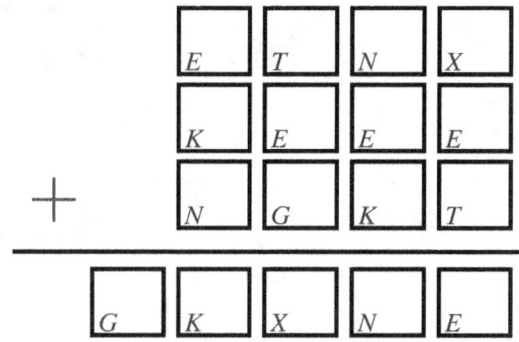

Problem No. 53:

$$
\begin{array}{r}
B G Y \\
B E Y \\
+ \quad N E Y \\
\hline
N E B E
\end{array}
$$

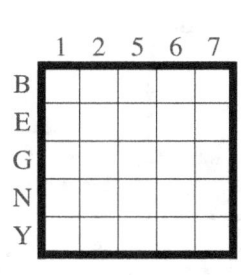

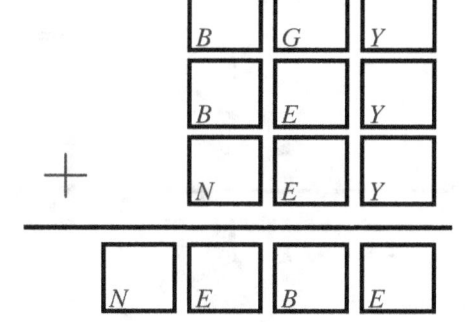

Problem No. 54:

```
  X Y K X
  X Y X A
+ K N K X
─────────
  H N T H K
```

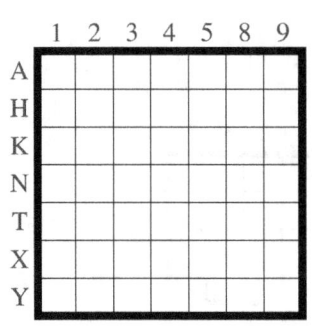

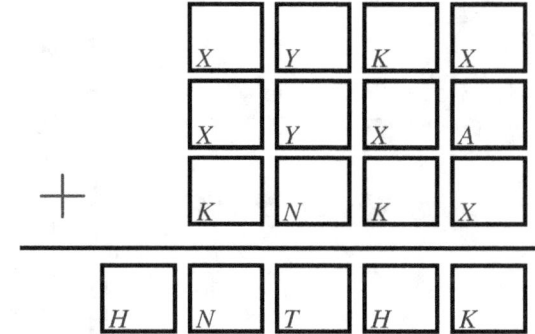

Problem No. 55:

```
  A G K E
  A G E X
+ G A E E
─────────
  X X X H G
```

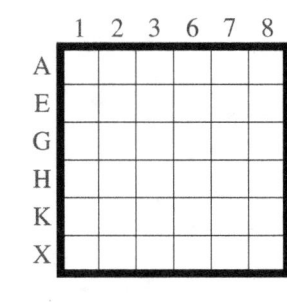

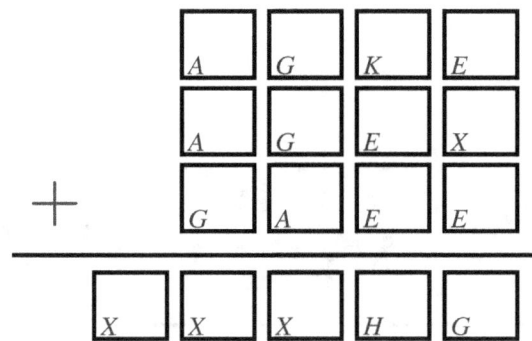

Problem No. 56:

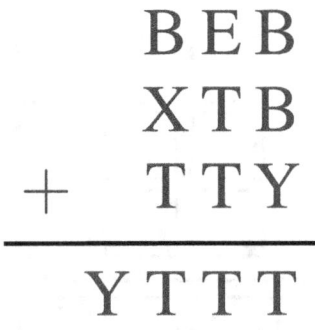

```
  B E B
  X T B
+ T T Y
───────
  Y T T T
```

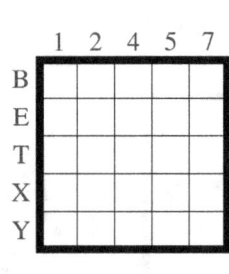

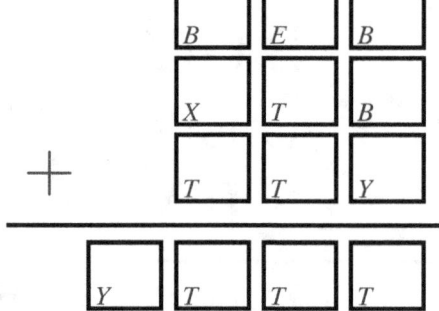

Problem No. 57:

```
    K Y E K
    K T T G
+   A K E E
_____
  G G A G B
```

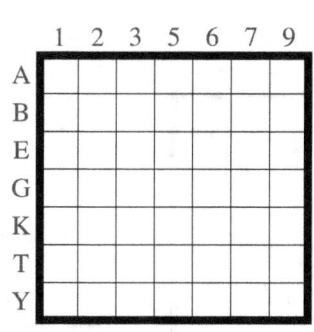
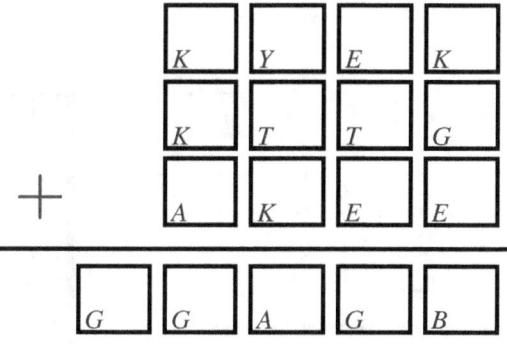

Problem No. 58:

```
      E H T
×     B K B
_____
      H Y N
    E E B E
    H Y N
_____
    K K N H N
```

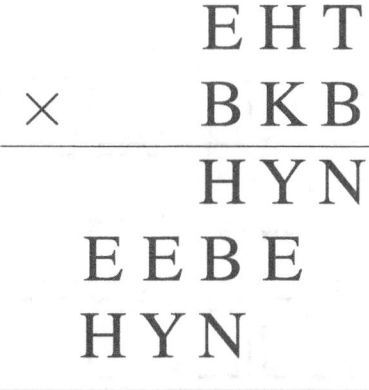
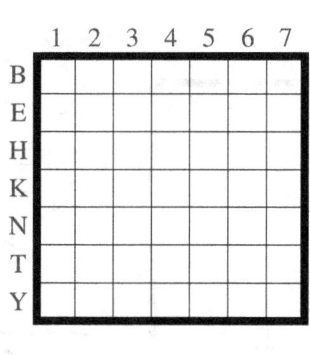
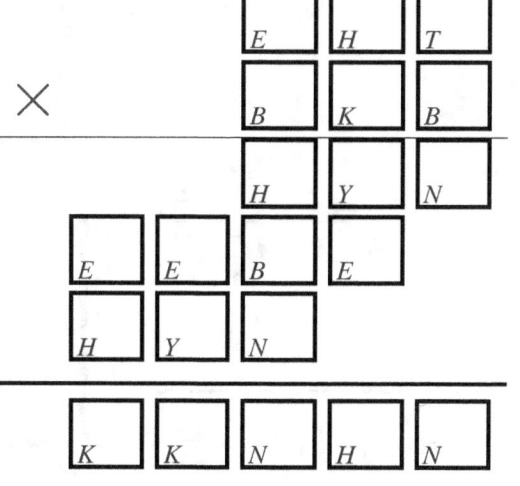

Problem No. 59:

```
    K E B
    B B A
+   K K K
_____
  E B H K
```

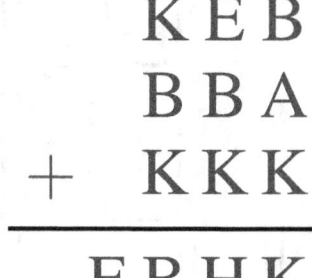
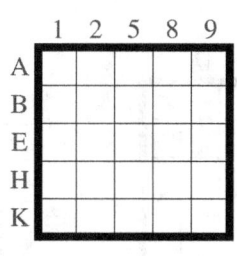

Problem No. 60:

```
      T T G
  ×     Y T
  ─────────────
      E E T
    G H Y Y
  ─────────────
    G X Y B T
```

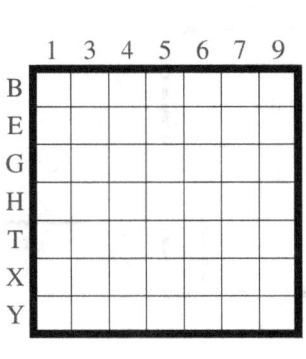

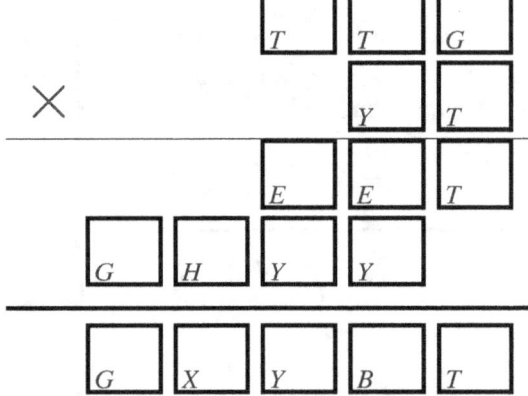

Problem No. 61:

```
      E H T
  ×     N X
  ─────────────
      B B K E
    T E Y N
  ─────────────
    H T K T E
```

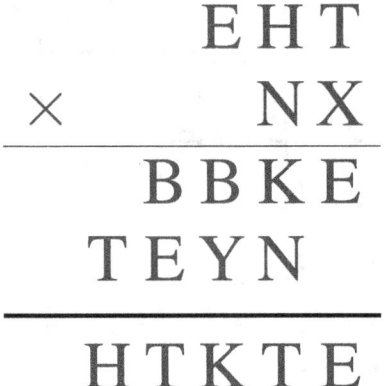

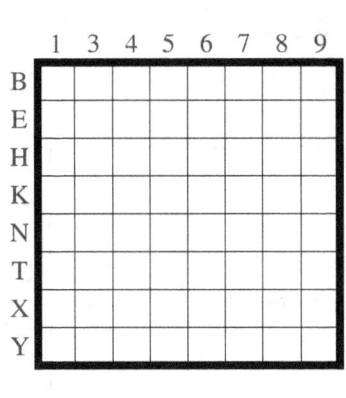

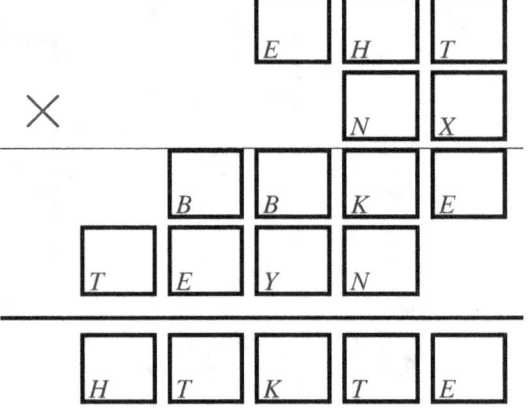

Problem No. 62:

```
      T H Y
  ×     T Y
  ─────────────
      T H Y
    Y X E T
  ─────────────
    Y N H X Y
```

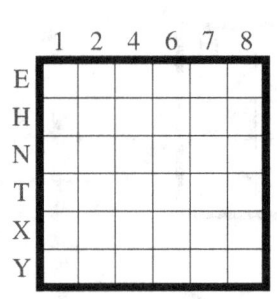

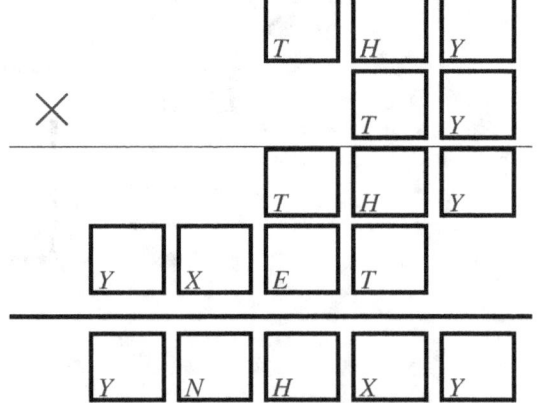

Problem No. 63:

```
  T K E H
  T K K E
+ Y H Y H
─────────
  E T N H T
```

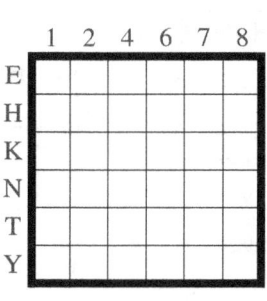

Problem No. 64:

```
    X G X
    H G H
  + X G H
  ─────────
    T H X G
```

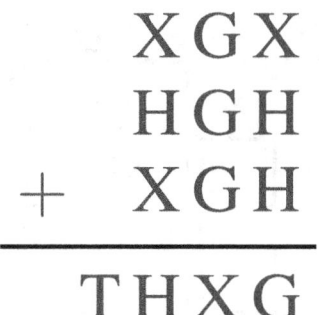
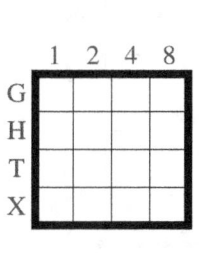
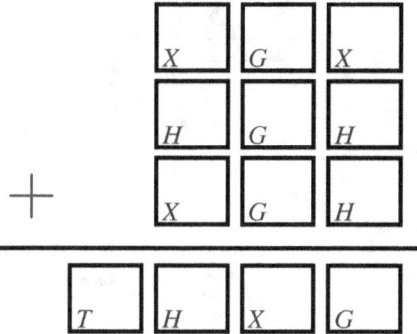

Problem No. 65:

```
    K H B
    N H N
  + G N G
  ─────────
    G H B N
```

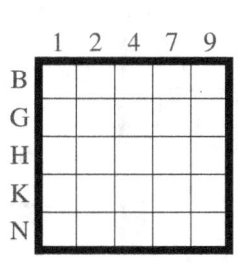
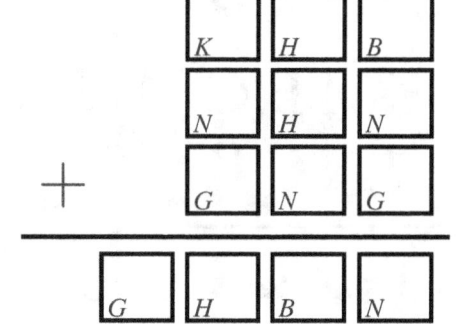

Problem No. 66:

```
  B T N
  T B N
+ N B N
-------
T Y B G
```

	1	2	3	6	9
B					
G					
N					
T					
Y					

Problem No. 67:

```
    T Y
×   N A
-------
  H T B
Y X B
-------
Y G T B
```

	0	1	2	5	6	7	8	9
A								
B								
G								
H								
N								
T								
X								
Y								

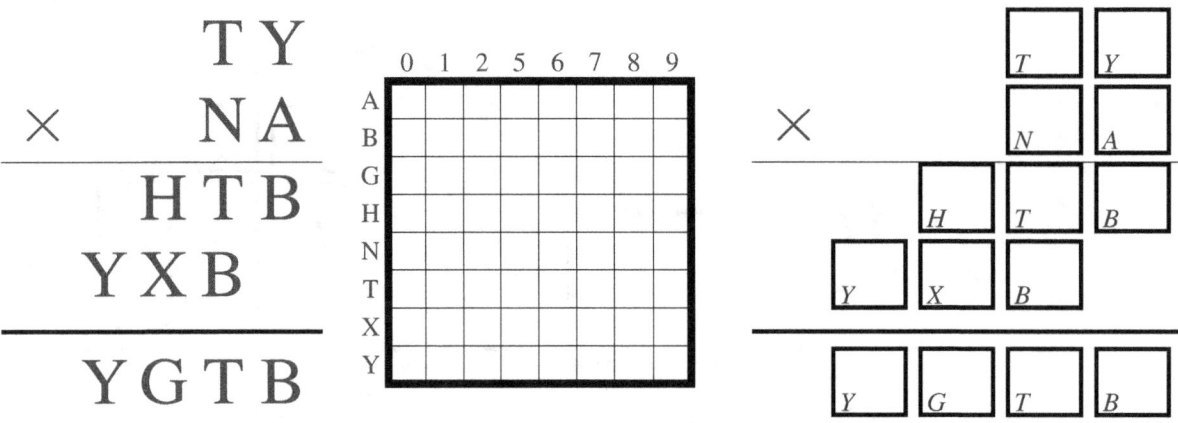

Problem No. 68:

```
  G E B H
  K H B G
+ T N G K
---------
E T T K K
```

	1	2	3	4	5	6	8
B							
E							
G							
H							
K							
N							
T							

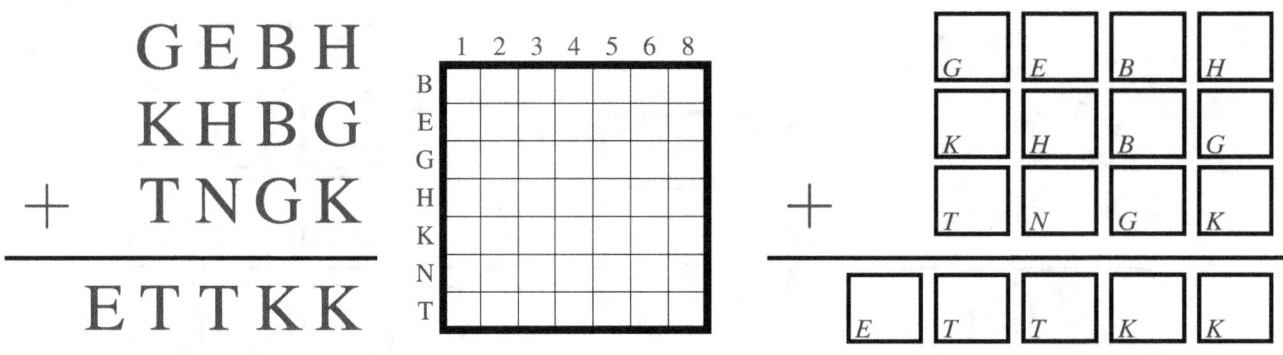

Problem No. 69:

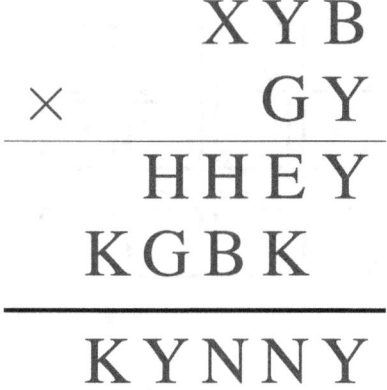

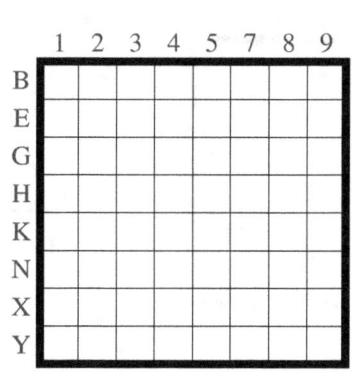

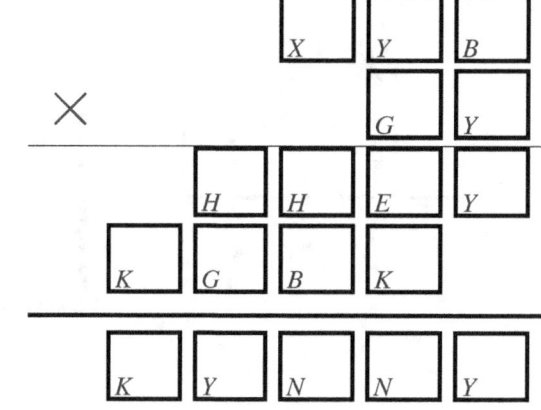

Problem No. 70:

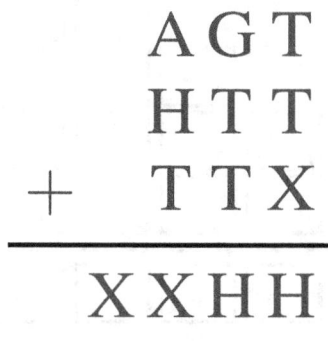

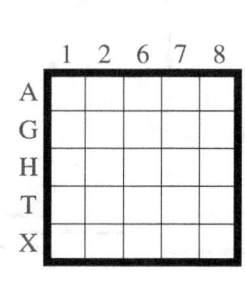

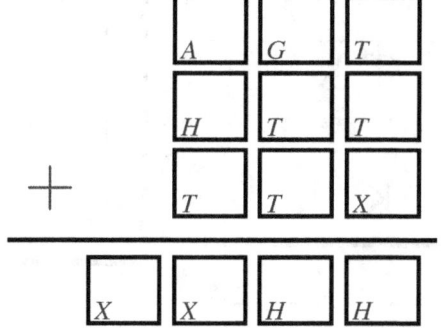

Problem No. 71:

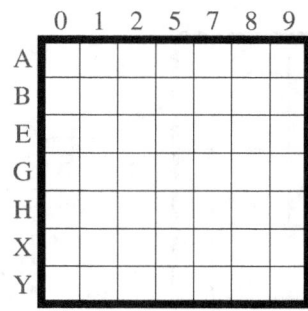

Problem No. 72:

$$
\begin{array}{r}
X H E E \\
+ \quad N G G A \\
\hline
K E N A G
\end{array}
$$

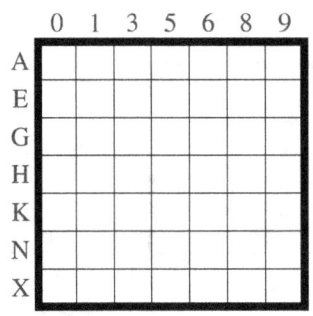

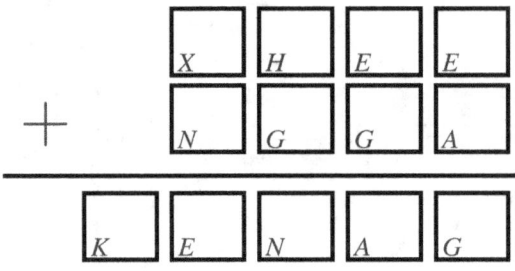

Problem No. 73:

$$
\begin{array}{r}
Y T X \\
\times \quad N X K \\
\hline
T X H E \\
H N Y T \\
K E K X \\
\hline
K X N X K E
\end{array}
$$

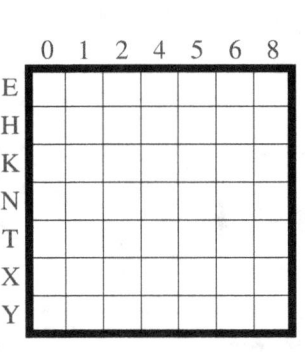

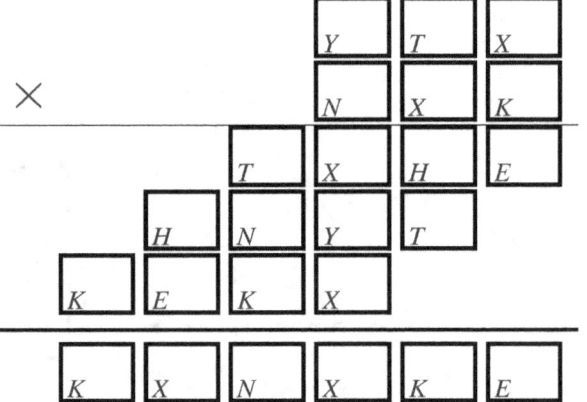

Problem No. 74:

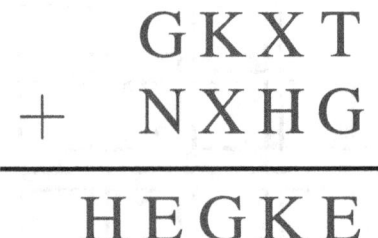

$$
\begin{array}{r}
G K X T \\
+ \quad N X H G \\
\hline
H E G K E
\end{array}
$$

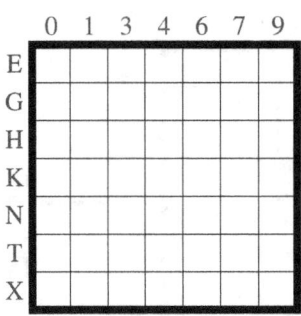

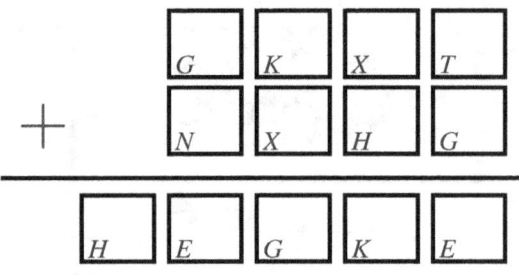

Problem No. 75:

```
  H H A
  K X A
+ G X H
-------
X G K G
```

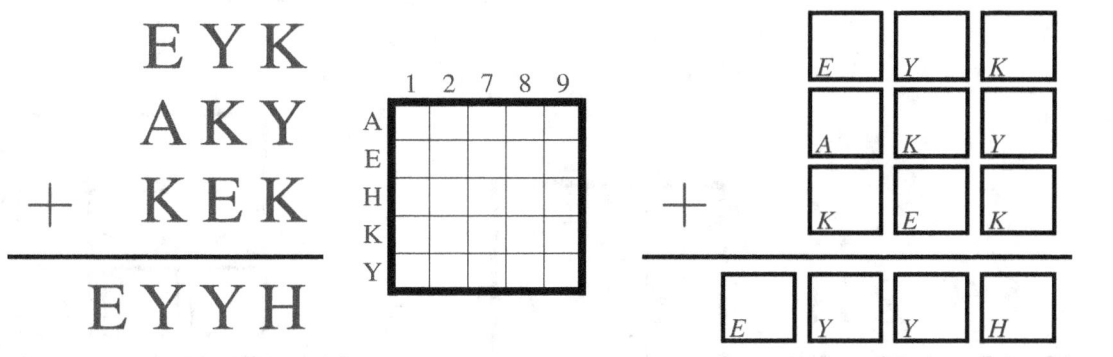

Problem No. 76:

```
  E Y K
  A K Y
+ K E K
-------
E Y Y H
```

Problem No. 77:

```
  H G A H
  A G T K
+ A B B K
---------
H A A B B
```

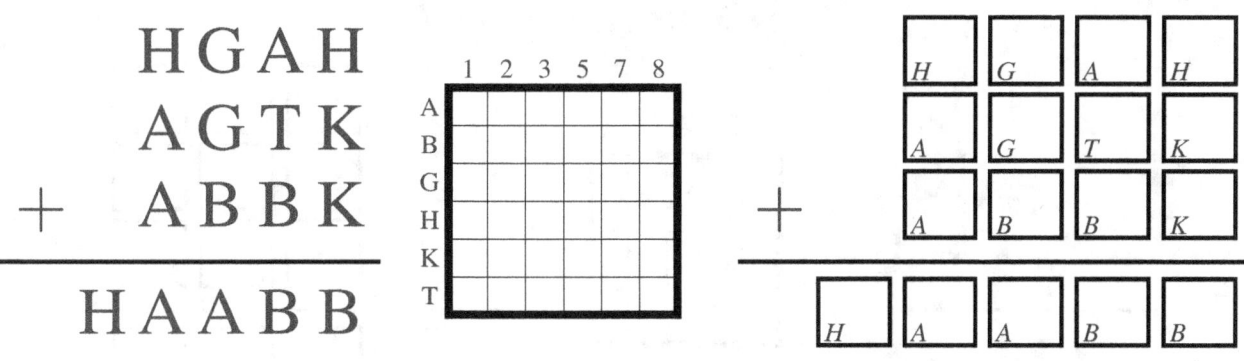

Problem No. 78:

```
      H B H
  ×   Y K T
  ─────────
      E K T
      Y B K
    T X Y
  ─────────
    E B E B T
```

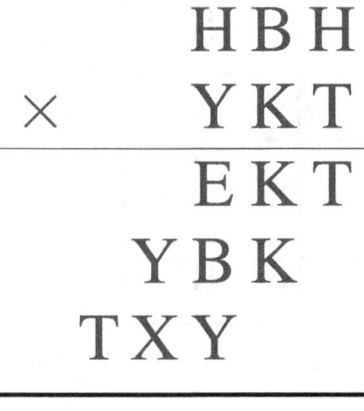
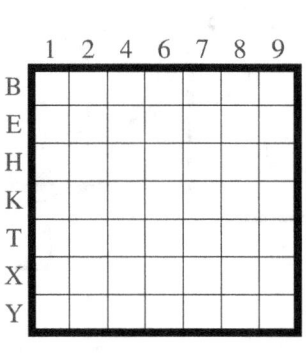
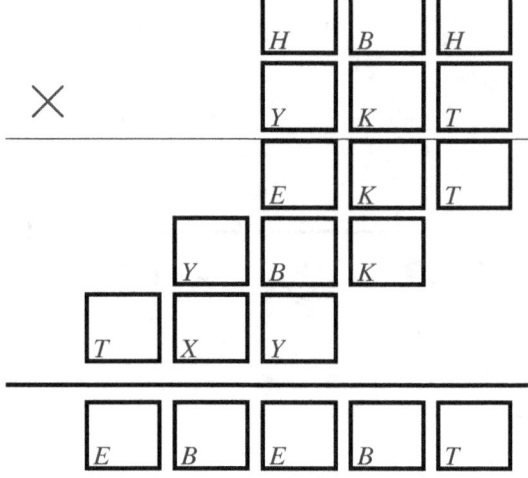

Problem No. 79:

```
      G Y B Y
      G H B X
  +   Y E B H
  ───────────
      Y X G X X
```

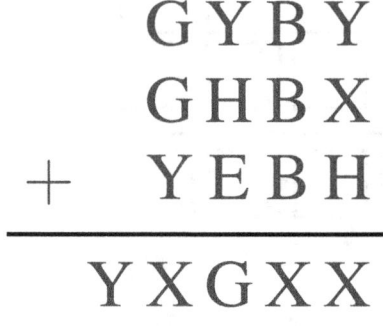
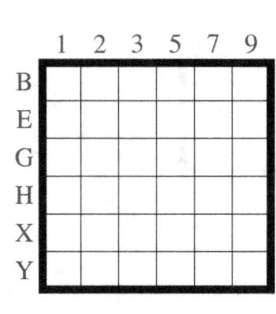
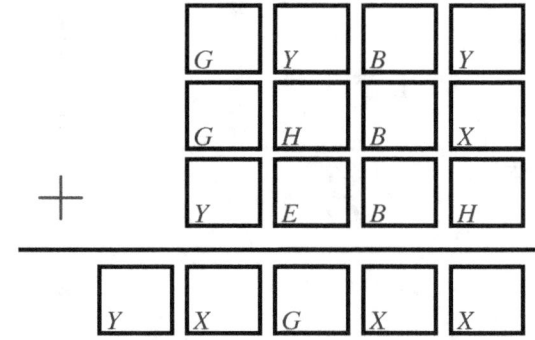

Problem No. 80:

```
      E E G
      T X E
  +   X E E
  ─────────
      T N G X
```

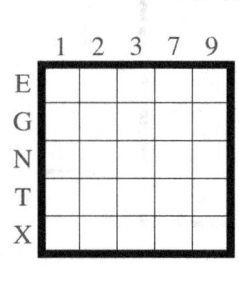
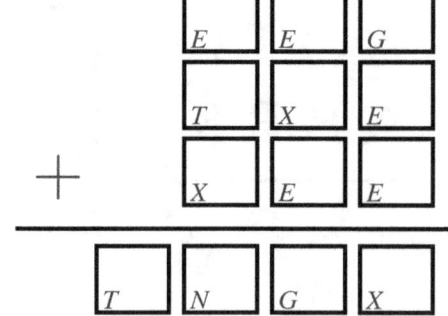

Problem No. 81:

$$
\begin{array}{r}
A\,G\,G \\
\times \quad G\,T \\
\hline
Y\,X\,K\,X \\
E\,H\,X\,X \\
\hline
G\,X\,H\,A\,X
\end{array}
$$

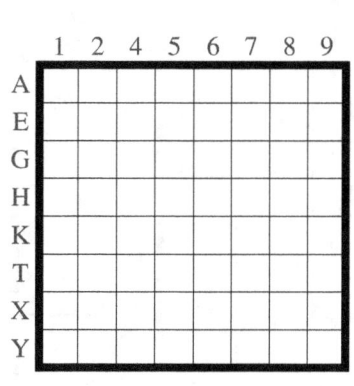

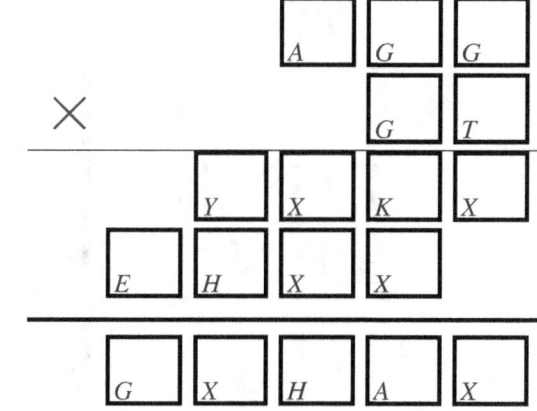

Problem No. 82:

$$
\begin{array}{r}
B\,Y\,Y\,N \\
+ \quad E\,X\,H\,A \\
\hline
N\,A\,B\,A\,X
\end{array}
$$

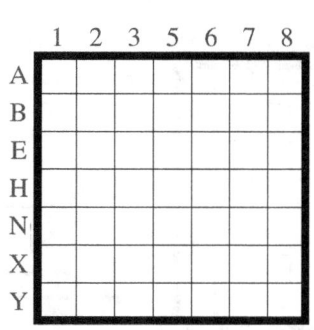

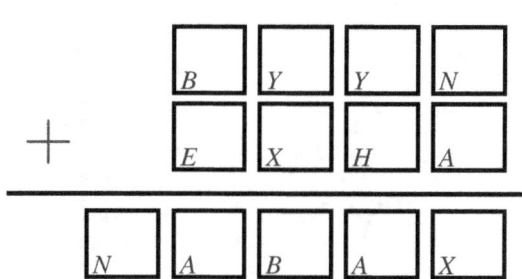

Problem No. 83:

$$
\begin{array}{r}
N\,H\,E\,N \\
+ \quad H\,G\,E\,A \\
\hline
K\,N\,A\,T\,E
\end{array}
$$

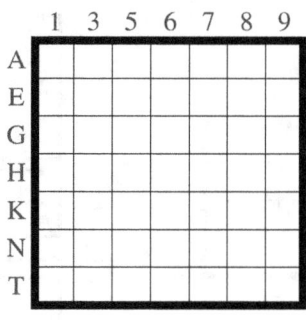

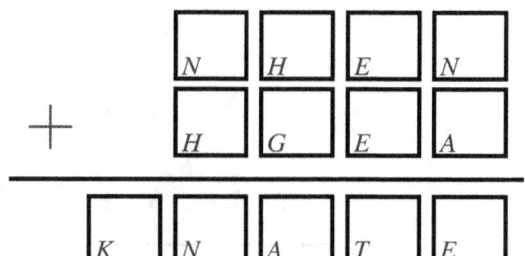

Problem No. 84:

```
   G B A
 ×   A B
 ---------
 T G Y B
 H B N
 ---------
 K K B B
```

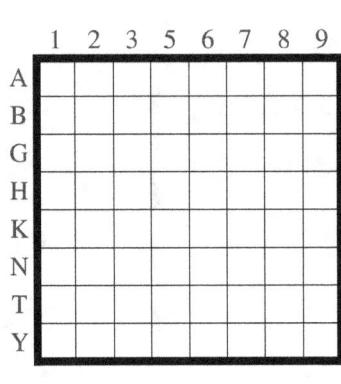
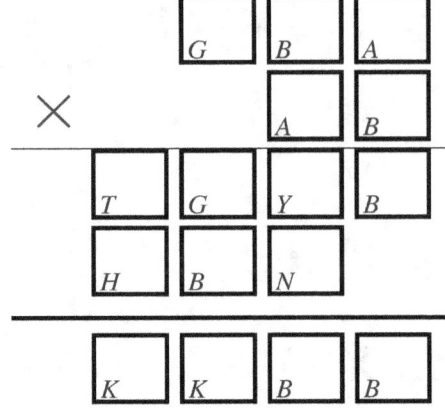

Problem No. 85:

```
   A A G
 ×   H E
 ---------
   A A G
 E Y N K
 ---------
 E K H G G
```

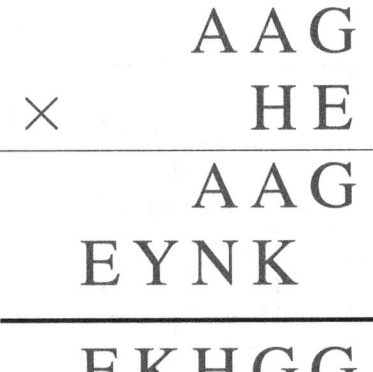
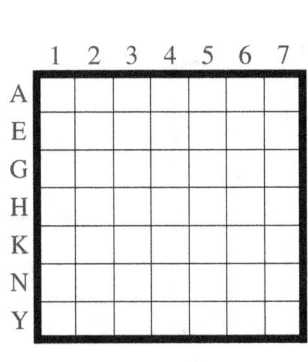
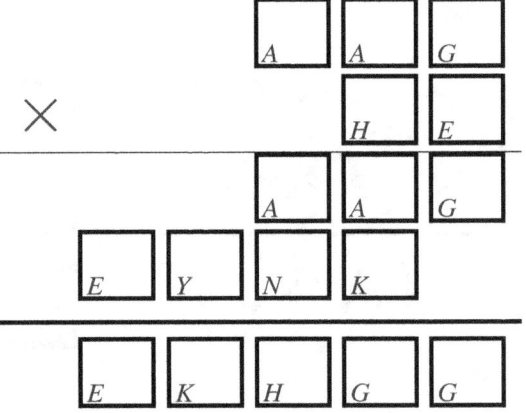

Problem No. 86:

```
   Y E A
   Y B Y
 + B A Y
 ---------
   E E B
```

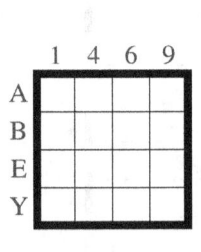
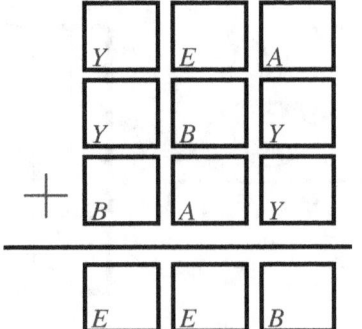

Problem No. 87:

```
    Y Y H
    X G X
  + T T Y
  ─────────
  H G Y Y
```

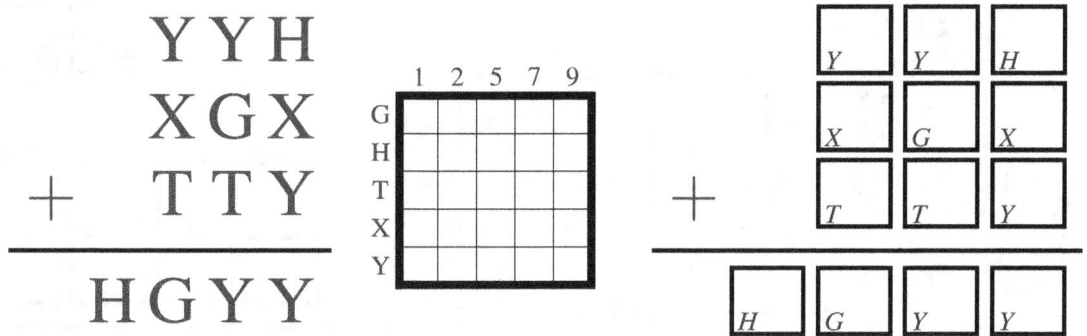

Problem No. 88:

```
      Y T K
  ×     B X
  ─────────
      Y T K
    K K G N
  ─────────
  K B B T K
```

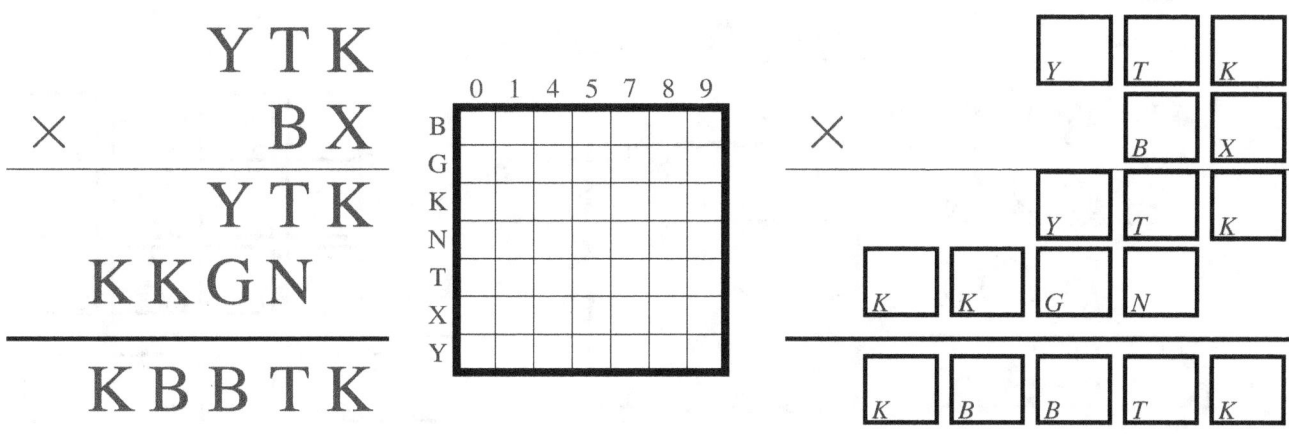

Problem No. 89:

```
    E X T
  ×   Y T
  ─────────
  A N H T
  E X T
  ─────────
  Y N N E T
```

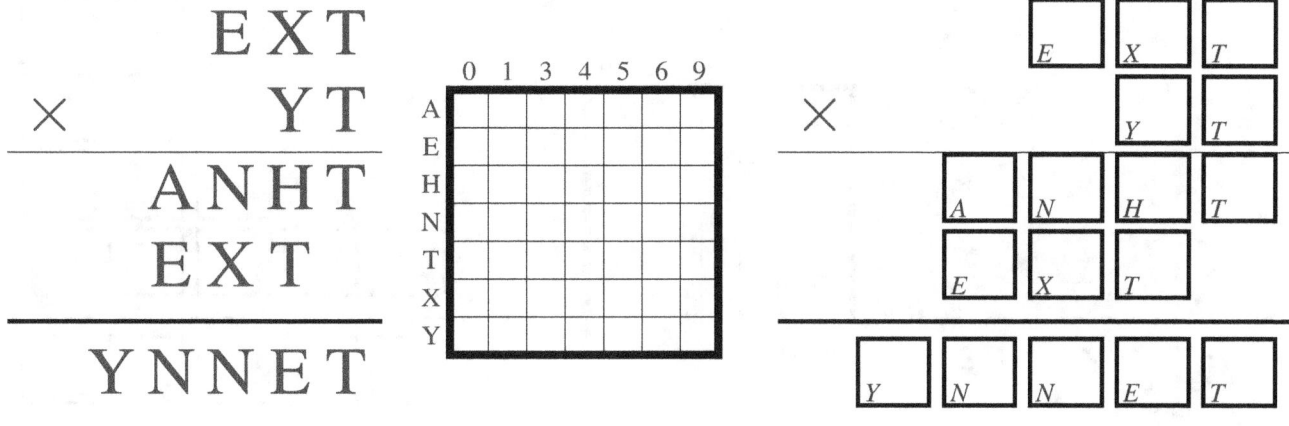

Problem No. 90:

```
    E B N
  ×   K K
  ─────────
  G H X T
  G H X T
  ─────────
  X B B N T
```

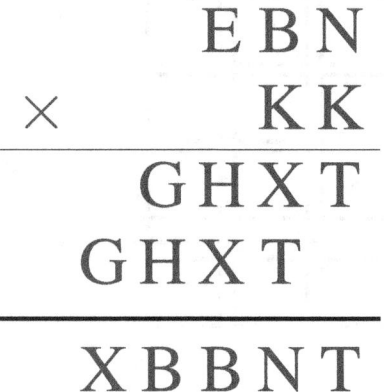
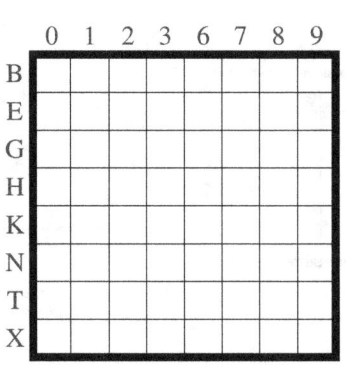
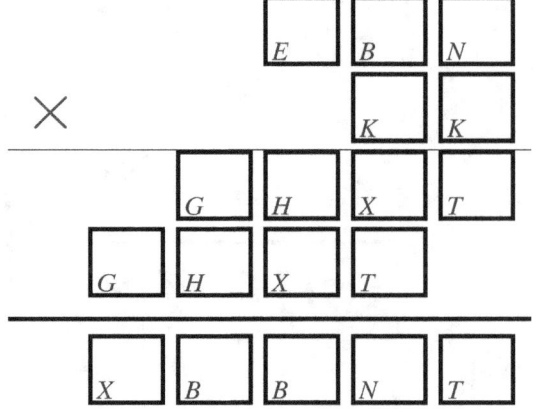

Problem No. 91:

```
  N B X H
  H G X N
+ B H N G
  ─────────
  K A G B A
```

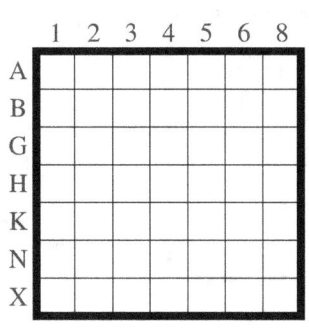
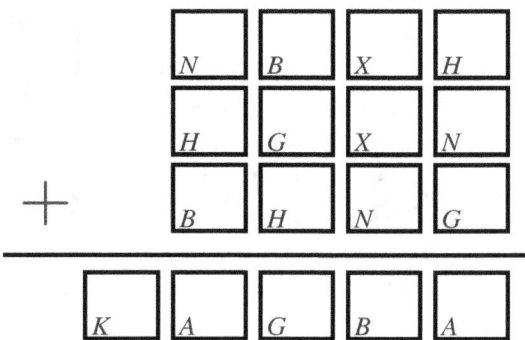

Problem No. 92:

```
    K Y B
  ×   N G
  ─────────
  K Y B
  G N E K
  ─────────
  G A G N B
```

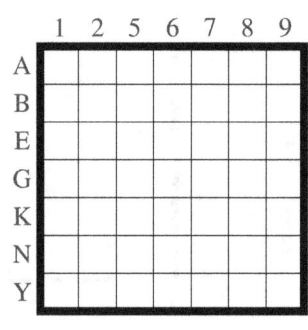
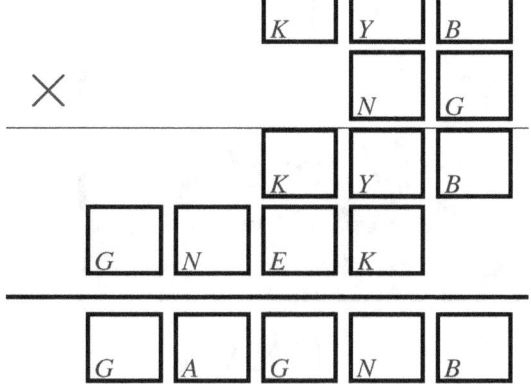

Problem No. 93:

```
    Y B A
    B G B
  + Y B A
  ---------
  X G Y A
```

	1	2	3	6	8
A					
B					
G					
X					
Y					

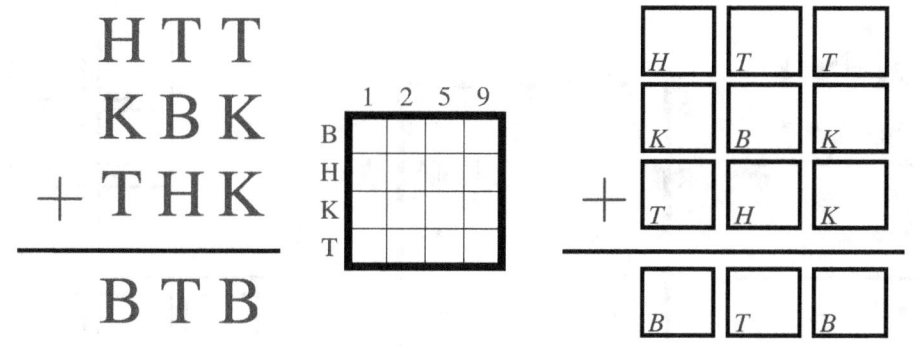

Problem No. 94:

```
    H T T
    K B K
  + T H K
  ---------
    B T B
```

	1	2	5	9
B				
H				
K				
T				

Problem No. 95:

```
    X E A T
    K B E E
  + T N X K
  -----------
  A X A N K
```

	1	2	3	4	6	8	9
A							
B							
E							
K							
N							
T							
X							

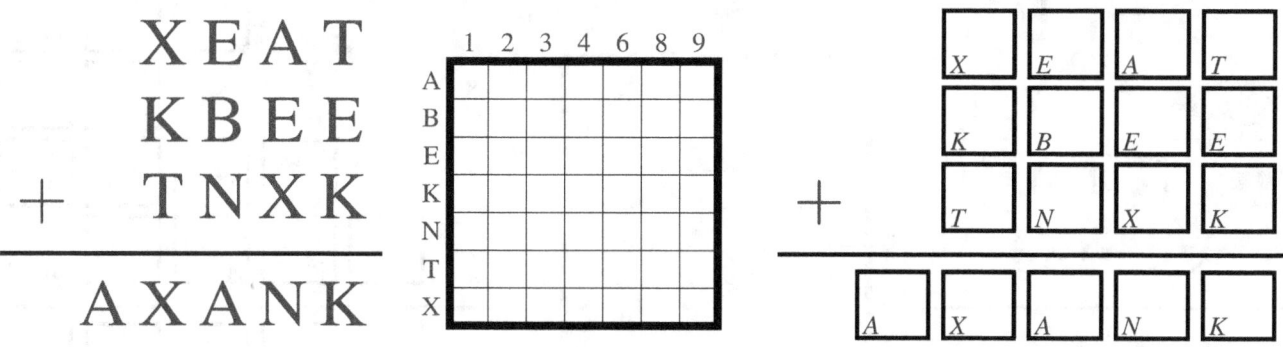

Problem No. 96:

```
    H H G
  ×   B K
  ─────────
    Y Y Y B
  K K K E
  ─────────
  K N N A B
```

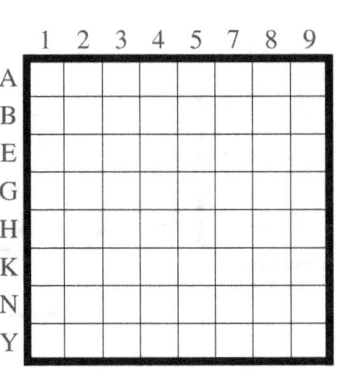

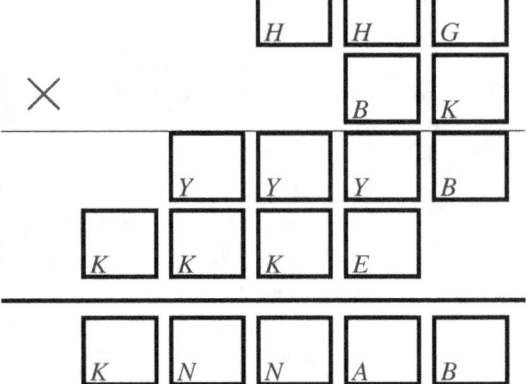

Problem No. 97:

```
    Y X N E
    T K Y X
  + N G X T
  ───────────
    E K Y X G
```

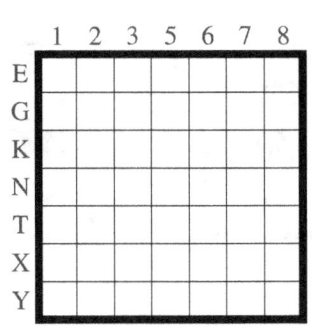

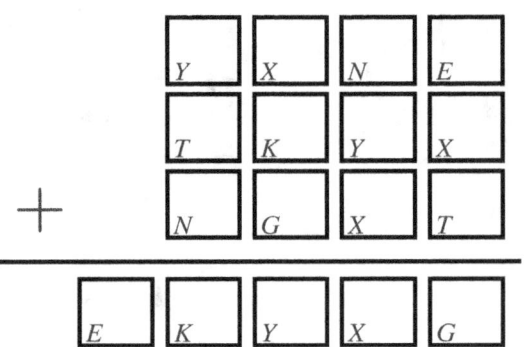

Problem No. 98:

```
    T E K
  ×   Y T
  ─────────
    X T Y E
  A T A G
  ─────────
  A B B A E
```

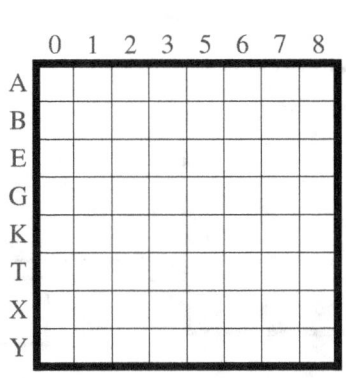

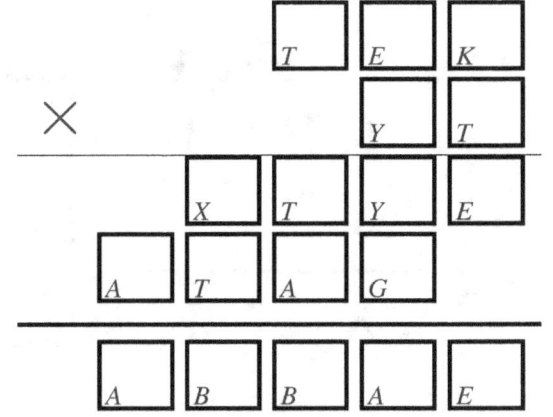

Problem No. 99:

```
    G B A
  ×   B A
  ─────────
  E E N K
K N G N
─────────────
K X X A K
```

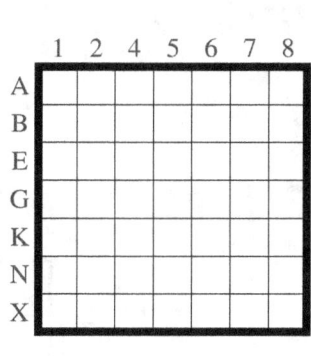
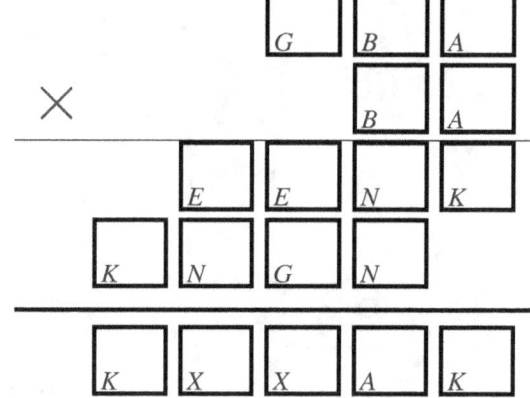

Problem No. 100:

```
    T Y T
  ×   B X
  ─────────
  T Y T
G T E K
─────────────
B E E B T
```

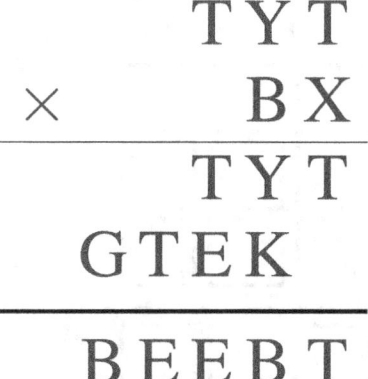
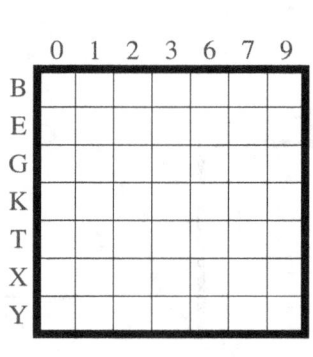
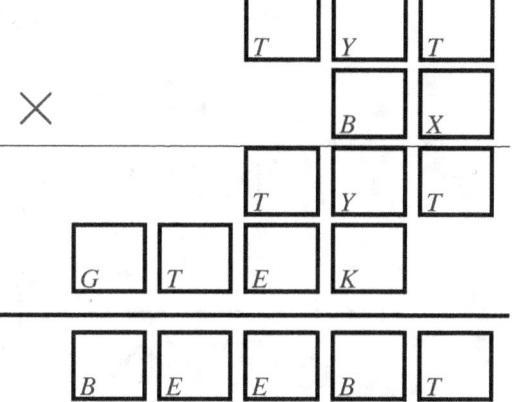

Problem No. 101:

```
    Y X E
  ×   K K
  ─────────
  G Y B Y
G Y B Y
─────────────
G K E E Y
```

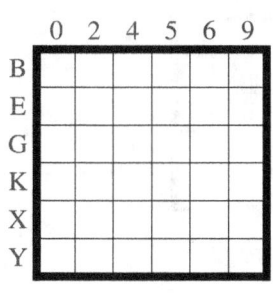
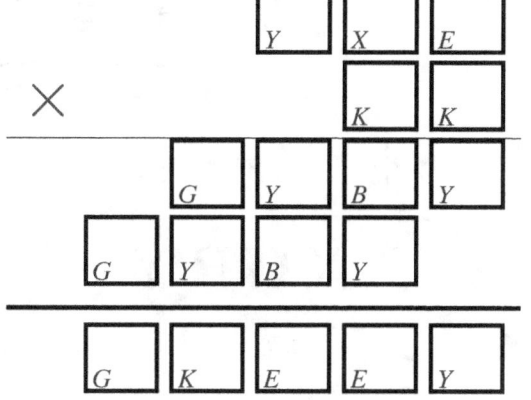

Problem No. 102:

```
        X B B
  ×     Y K B
  ─────────────
        X B B
      B X K K
    K X Y Y
  ─────────────
    K Y K X N B
```

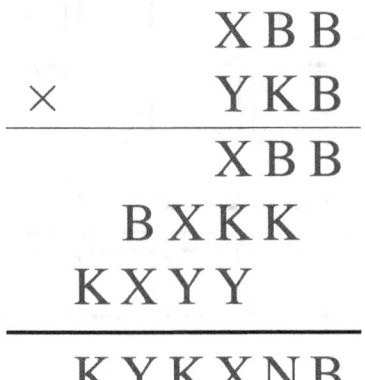

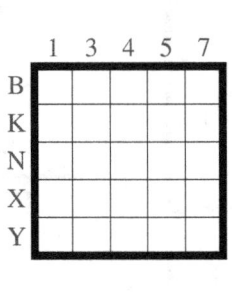

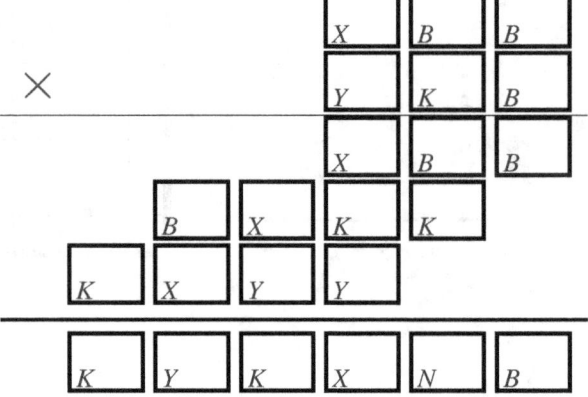

Problem No. 103:

```
      Y G Y
      N N Y
  +   N T T
  ───────────
      K T K N
```

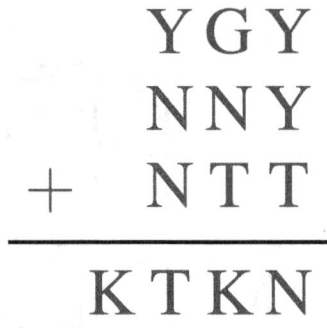

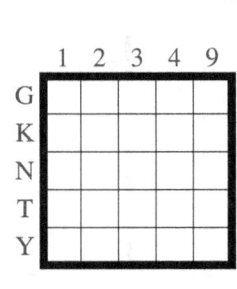

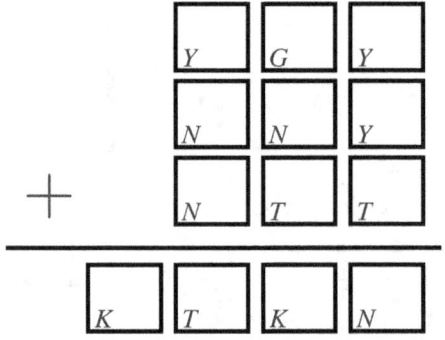

Problem No. 104:

```
      B X B
      T T G
  +   G T B
  ───────────
      X G B Y
```

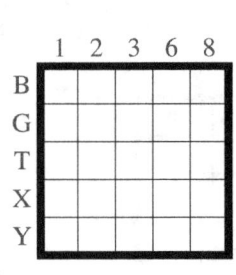

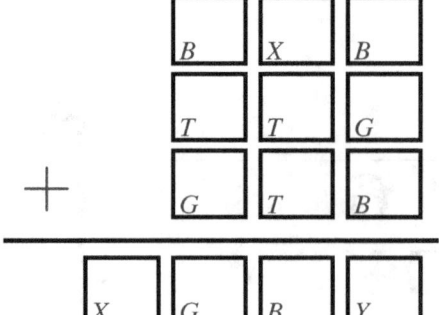

Problem No. 105:

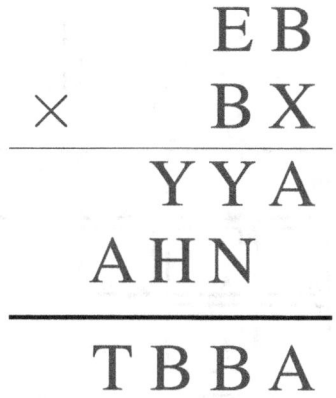

$$
\begin{array}{r}
E\,B \\
\times \quad B\,X \\
\hline
Y\,Y\,A \\
A\,H\,N \\
\hline
T\,B\,B\,A
\end{array}
$$

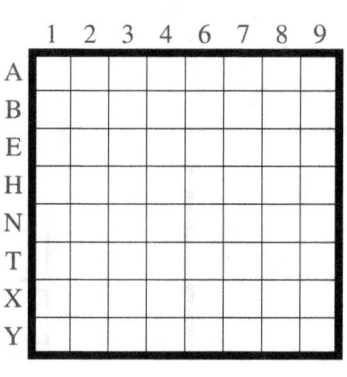

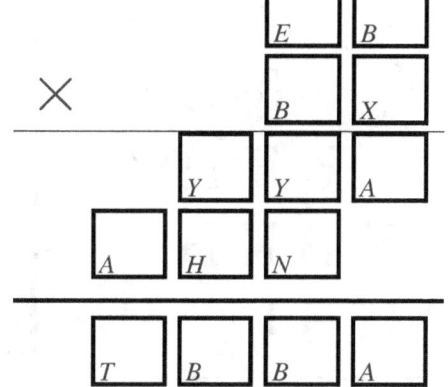

Problem No. 106:

$$
\begin{array}{r}
X\,B\,X \\
\times \quad K\,E \\
\hline
Y\,Y\,K\,K \\
A\,Y\,X\,Y \\
\hline
A\,G\,A\,E\,K
\end{array}
$$

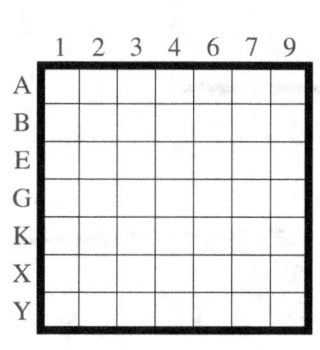

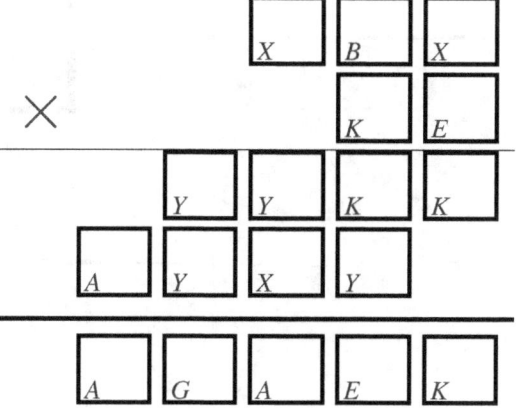

Problem No. 107:

$$\begin{array}{r} X\,X\,N \\ \times\quad T\,B\,H \\ \hline K\,T\,H\,X \\ K\,B\,K\,H \\ H\,B\,K \\ \hline B\,N\,H\,X\,X \end{array}$$

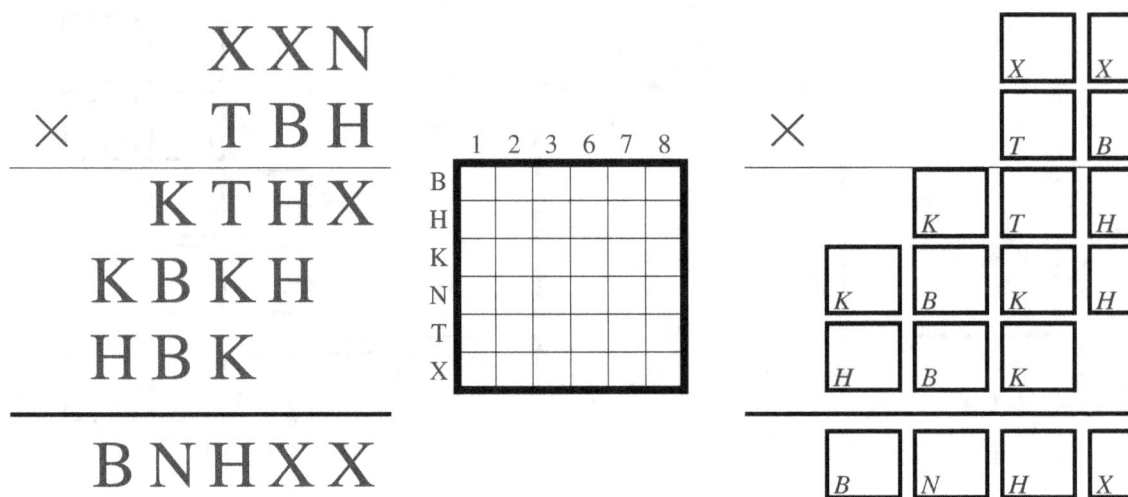

Problem No. 108:

$$\begin{array}{r} X\,G\,Y\,B \\ X\,B\,T\,T \\ +\quad B\,B\,A\,B \\ \hline G\,X\,Y\,K\,Y \end{array}$$

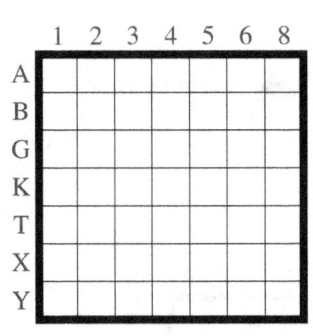
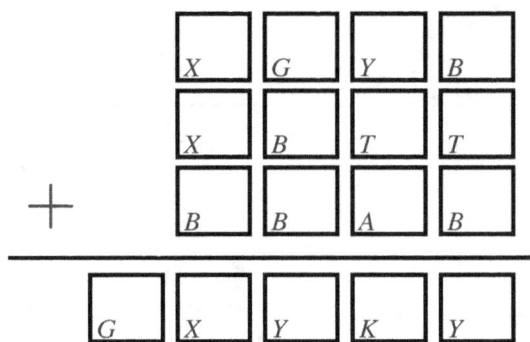

Problem No. 109:

$$\begin{array}{r} A\,E \\ \times\quad G\,X \\ \hline T\,B\,K \\ X\,H\,B \\ \hline X\,G\,X\,K \end{array}$$

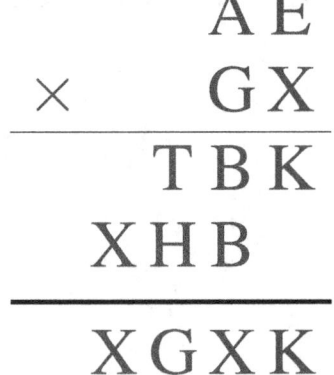
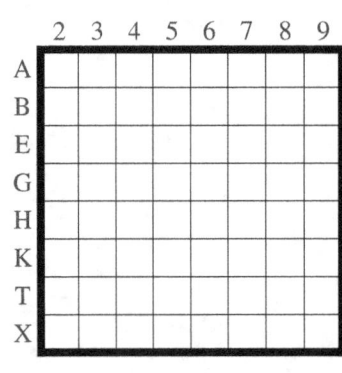
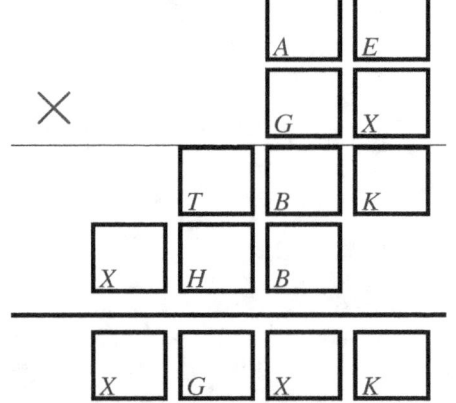

Problem No. 110:

```
  H X G H
  K A A B
+ B K K A
─────────
  E H A X B
```

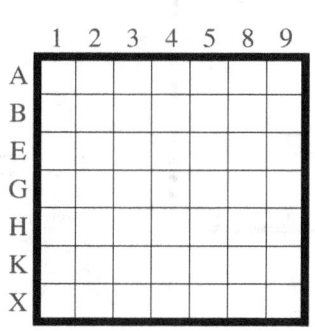
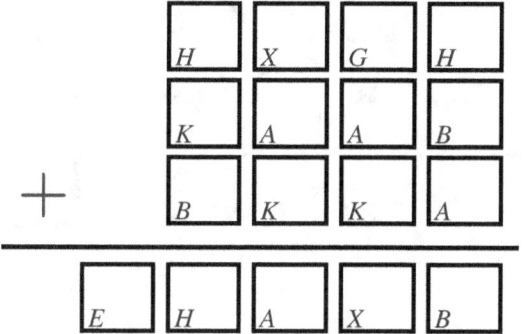

Problem No. 111:

```
    K T G
    T T G
+   T X H
─────────
  K T H T
```

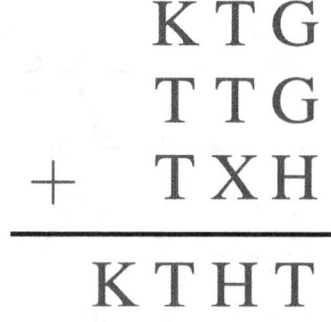
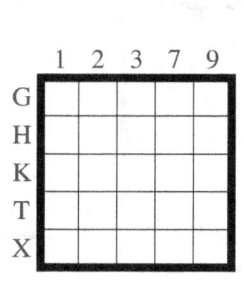
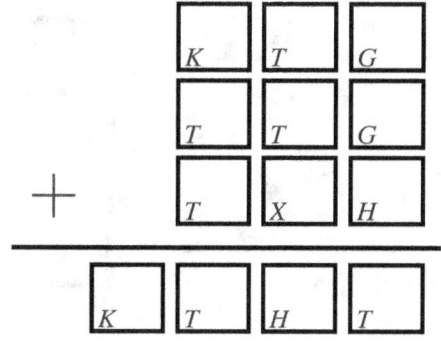

Problem No. 112:

```
      K E H
  ×   X N A
  ─────────
    Y N E X
    K E H
  N X K A
  ─────────
  N X H K X X
```

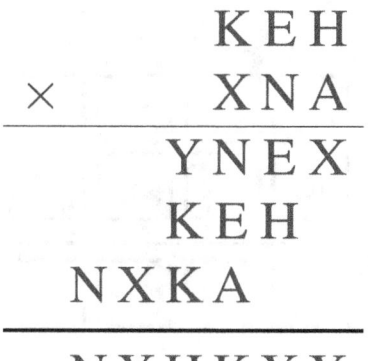
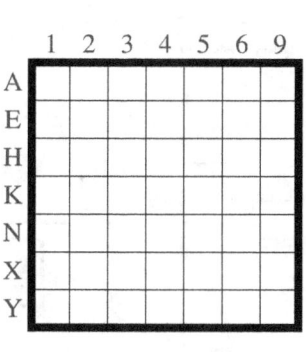
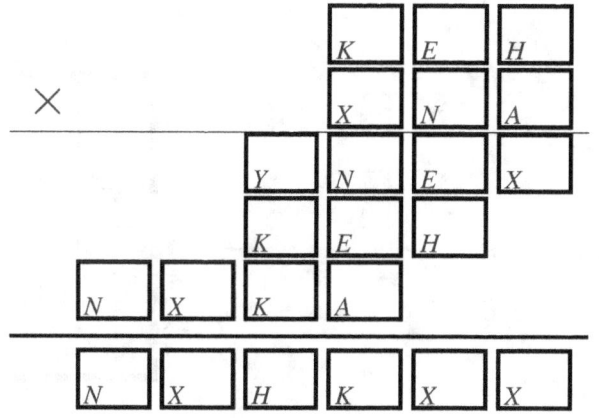

Problem No. 113:

```
  B G A G
  G B X K
+ K N H A
─────────
  G H G N N
```

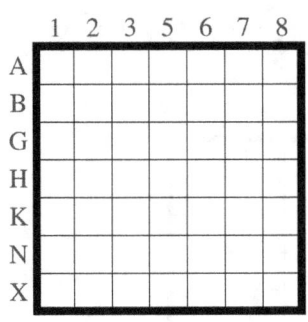
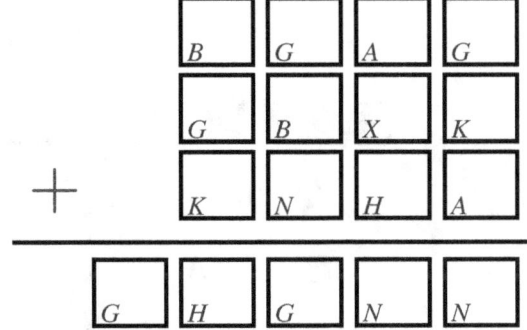

Problem No. 114:

```
    G H E
  ×   K E
─────────
    X T A
  T K E X
─────────
  T H T A A
```

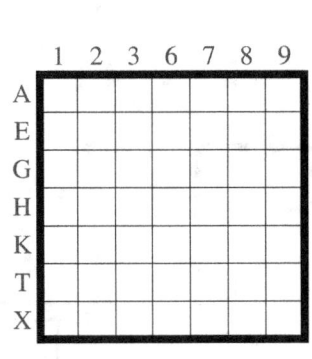
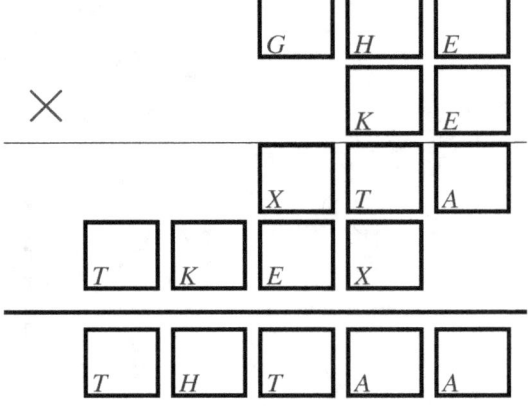

Problem No. 115:

```
  X H N B
+ X E X H
─────────
  Y Y T Y A
```

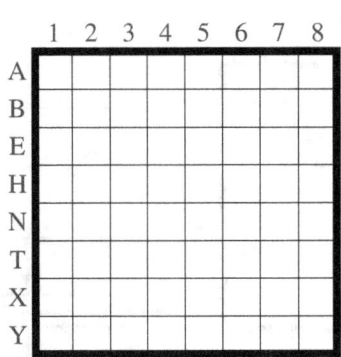
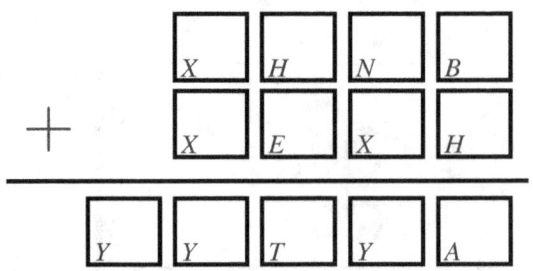

Problem No. 116:

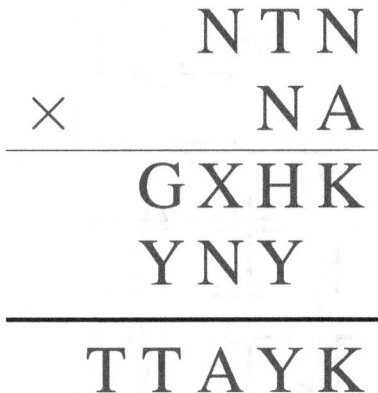

```
    N T N
  ×   N A
  ─────────
  G X H K
  Y N Y
  ─────────
  T T A Y K
```

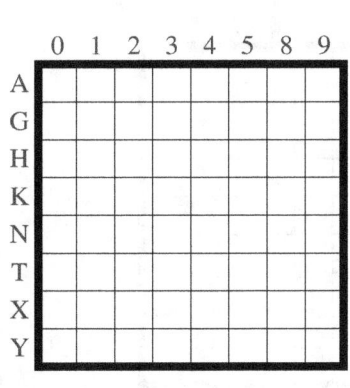

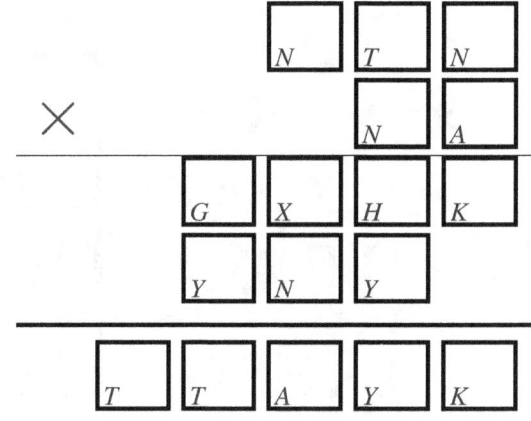

Problem No. 117:

```
    G N B
  ×   A X
  ─────────
  N T E Y
  N X N G
  ─────────
  G Y G G Y
```

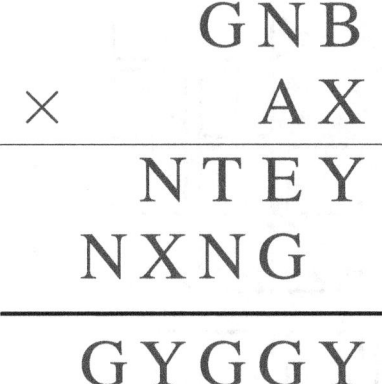

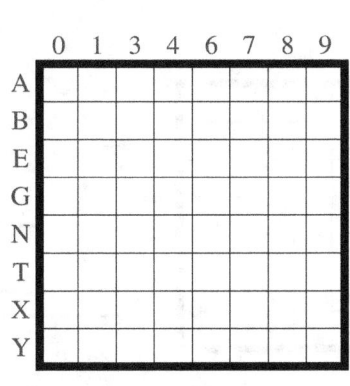

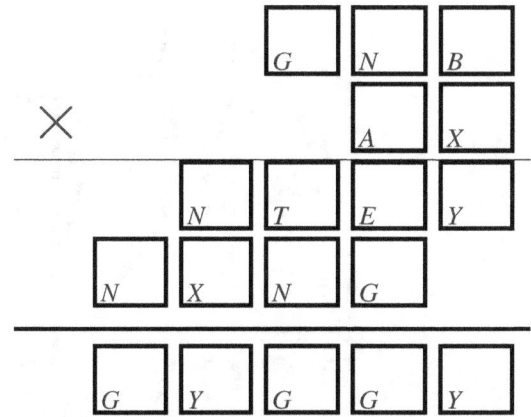

Problem No. 118:

```
    X B B
    B T H
  +   B T H
  ─────────
  H A T X
```

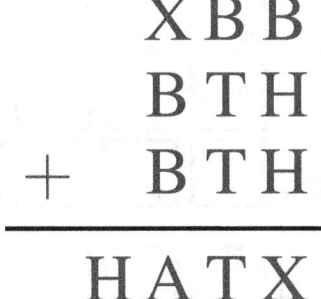

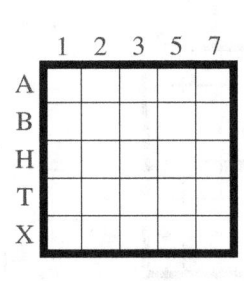

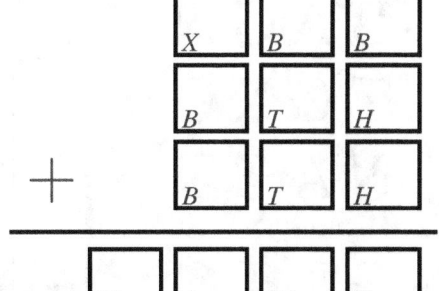

Problem No. 119:

```
    T K
  ×   K T
  ───────
    X N G
  H E A
  ───────
  B K T G
```

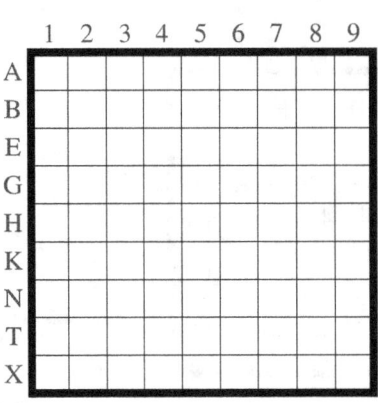
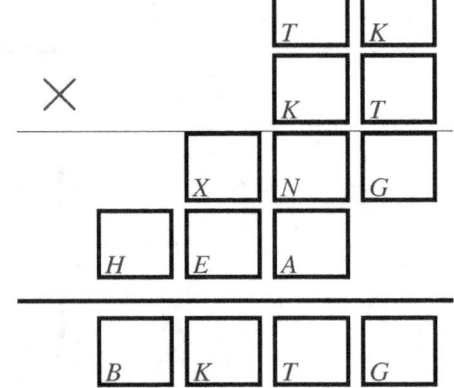

Problem No. 120:

```
    A Y
  ×   Y B
  ───────
    T N Y
  B E Y
  ───────
  B G G Y
```

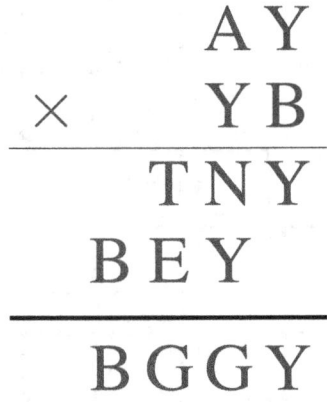
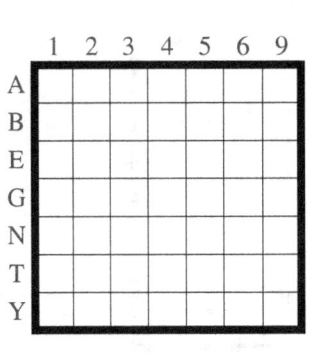
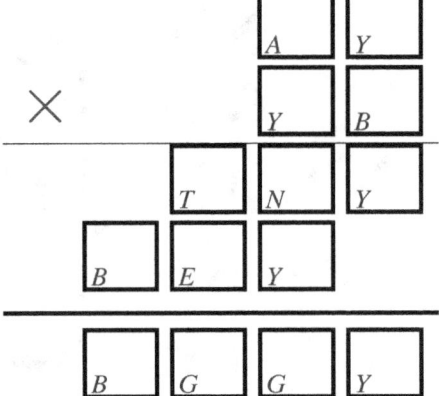

Problem No. 121:

```
  B A G N
  T G X G
  + N A G T
  ─────────
  N G T T G
```

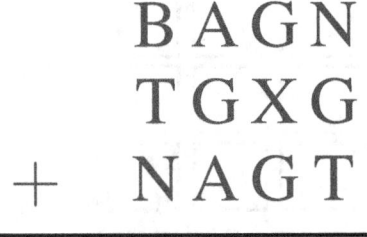
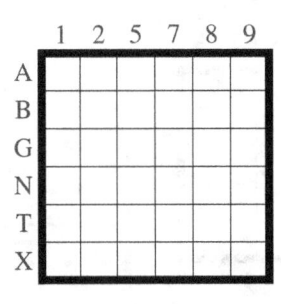
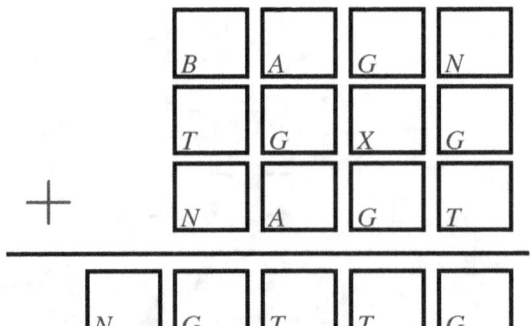

Problem No. 122:

```
    T T B
  ×   T H
  ─────────
  Y T B Y
  B T B X
  ─────────
  T Y B A Y
```

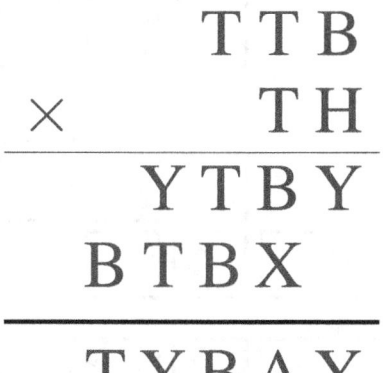
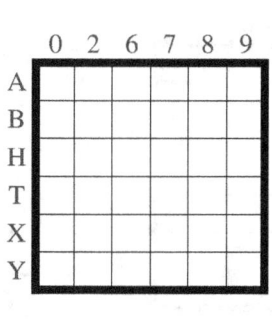
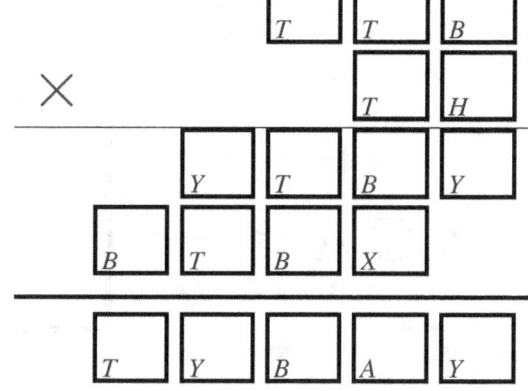

Problem No. 123:

```
    E A E
    E A E
  + E E N
  ─────────
    N T E
```

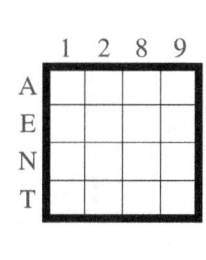
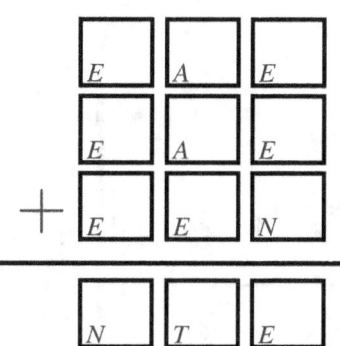

Problem No. 124:

```
      E E Y
  ×   T X E
  ─────────
    X A G E
    E E Y
  X T T G
  ─────────
  X E K K E E
```

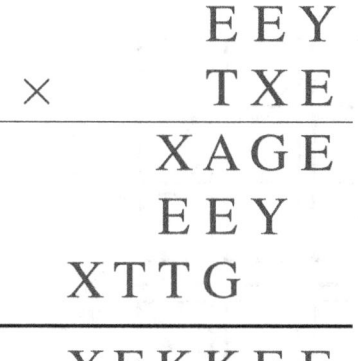
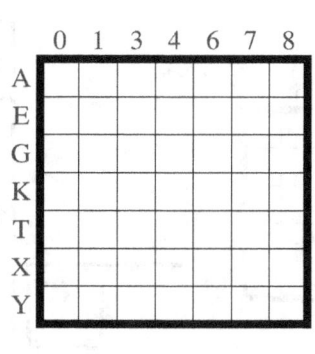
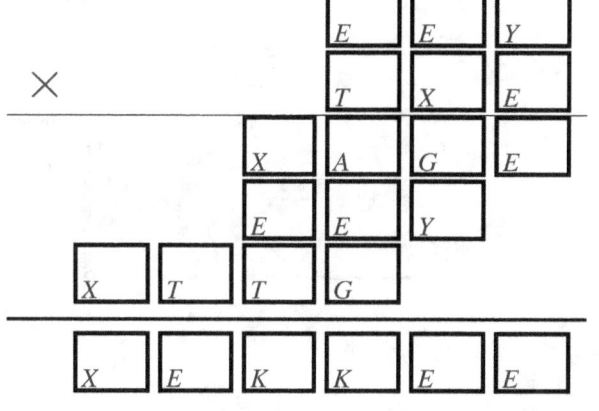

Problem No. 125:

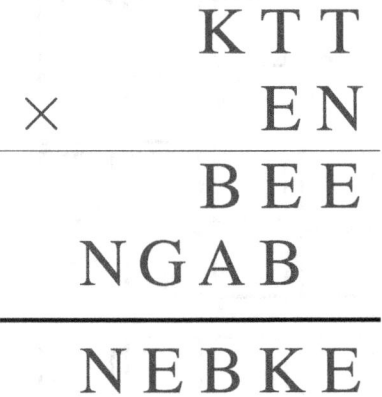

```
    K T T
  ×   E N
  ———————
    B E E
  N G A B
  ———————
  N E B K E
```

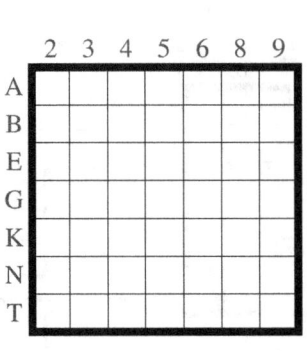

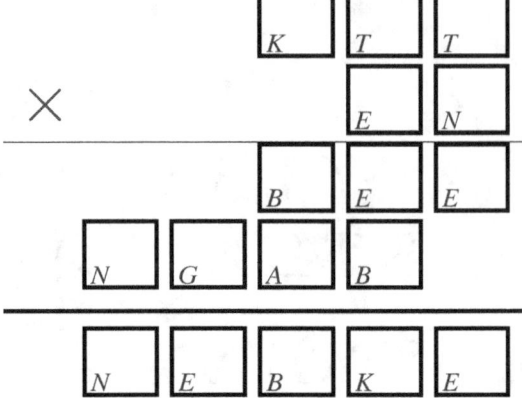

Problem No. 126:

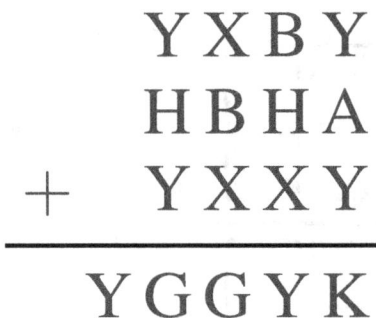

```
  Y X B Y
  H B H A
  + Y X X Y
  ———————
  Y G G Y K
```

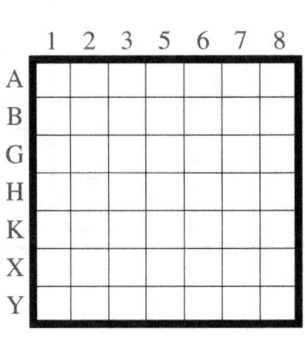

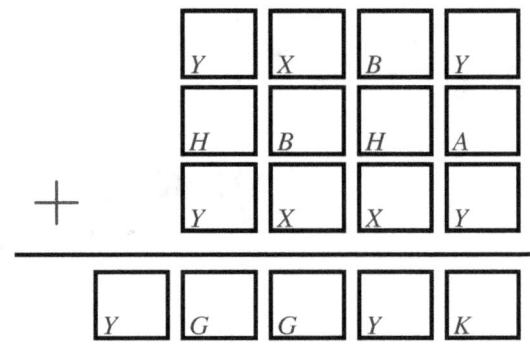

Problem No. 127:

```
    T N N
    Y T N
  + E A N
  ———————
    T Y Y A
```

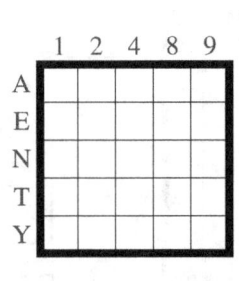

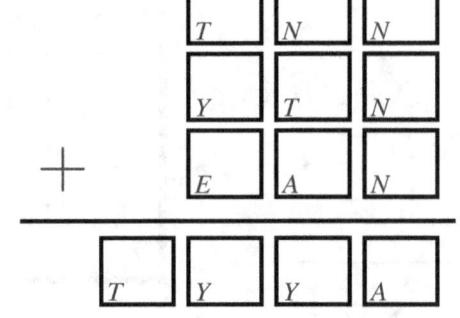

Problem No. 128:

```
    E E T
    T T B
  + E B K
  ─────────
    T T K B
```

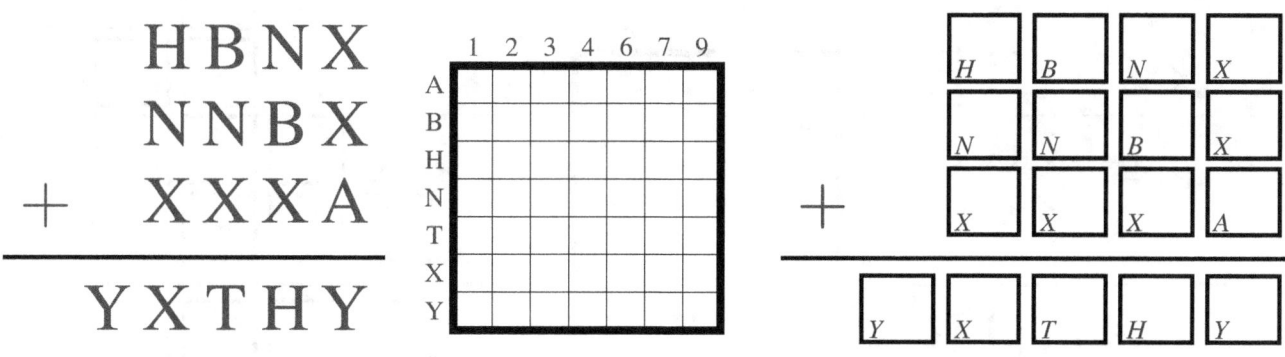

Problem No. 129:

```
    H B N X
    N N B X
  + X X X A
  ─────────
    Y X T H Y
```

Problem No. 130:

```
    T H E B
  + A G G G
  ─────────
    T T B G X
```

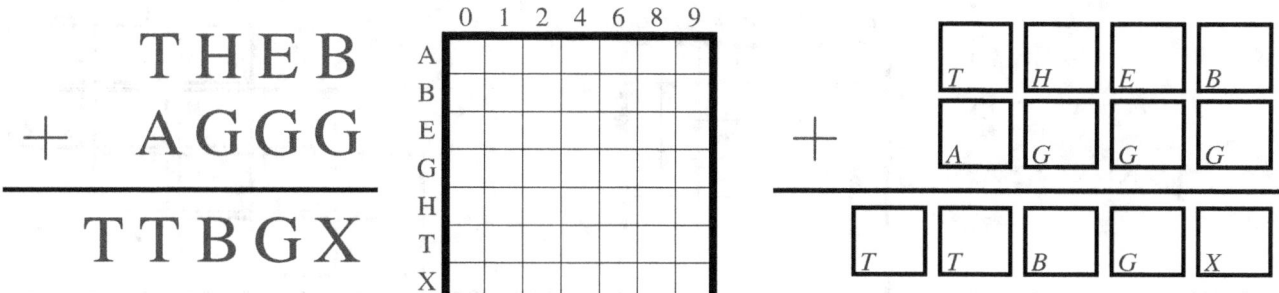

Problem No. 131:

```
  N Y T H
  H A N K
+ K A T Y
─────────
  A B H K Y
```

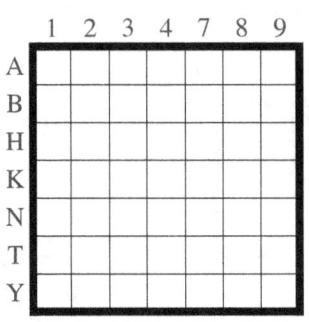
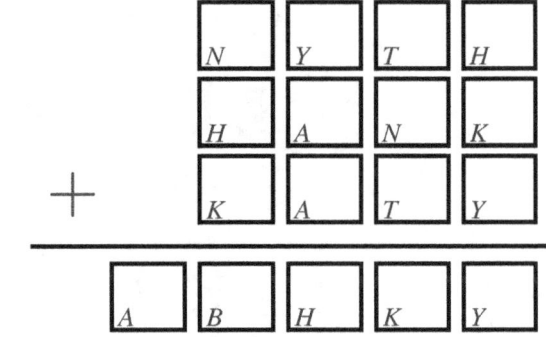

Problem No. 132:

```
      X G
  ×   H E
  ─────────
      B B E
    A K X
  ─────────
    H H H E
```

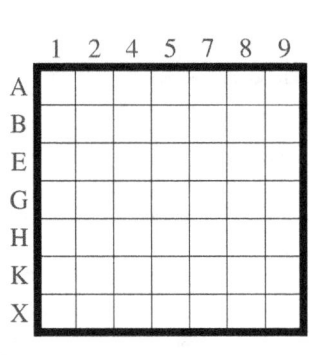
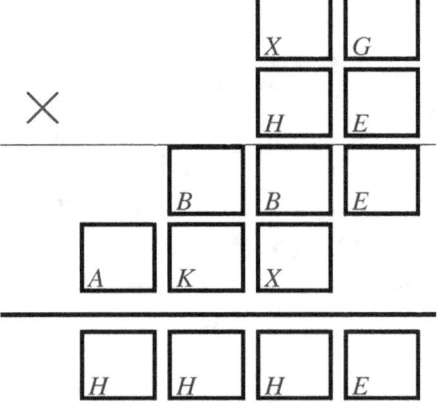

Problem No. 133:

```
        N N E
  ×     A E G
  ───────────
      N T T B
    K N N G
    N N E
  ───────────
  A A G B G B
```

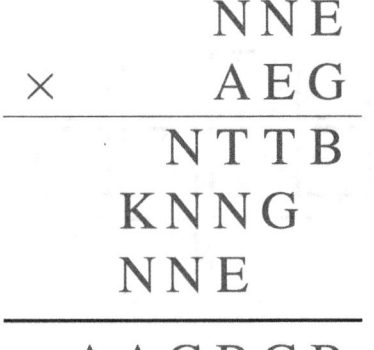
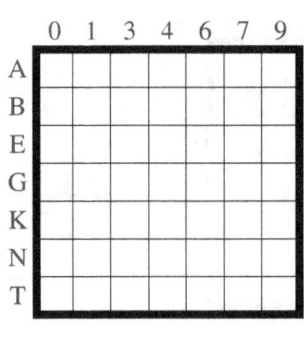
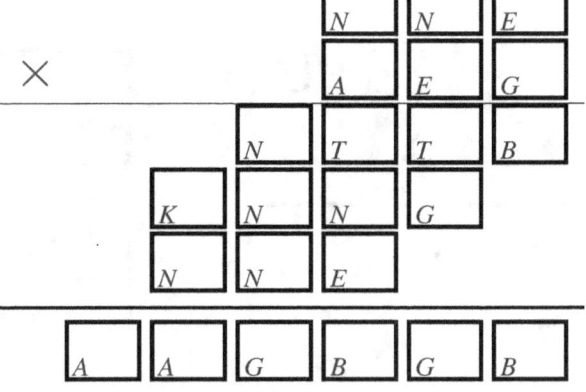

Problem No. 134:

```
   K G B Y
 + N K X Y
 ─────────
   A K B A K
```

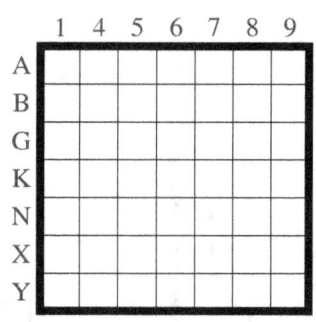

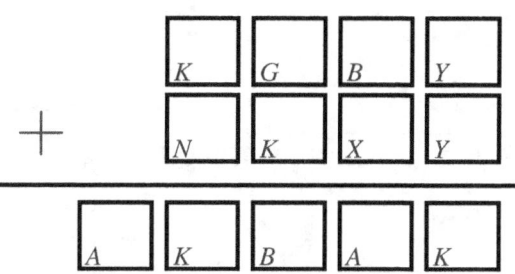

Problem No. 135:

```
   A H E E
   A H K G
 + K E E A
 ─────────
   B K G B G
```

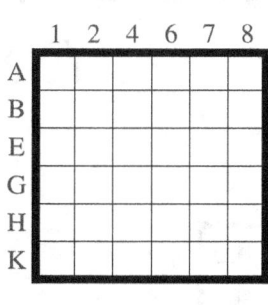

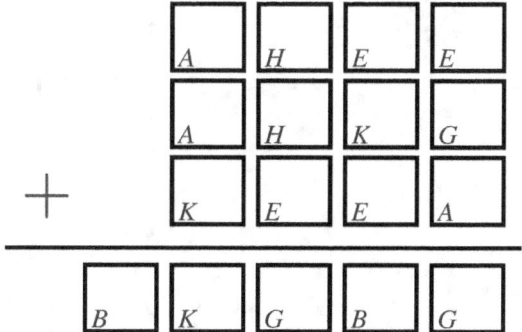

Problem No. 136:

```
     T N B
 ×     B X
 ─────────
   K N N N
   H T H K
 ─────────
   B K G B N
```

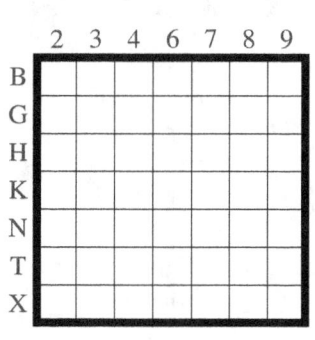

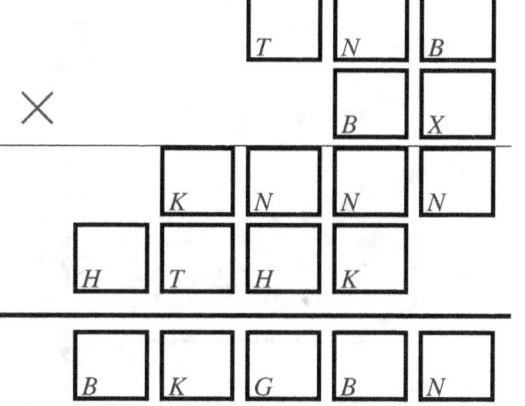

Problem No. 137:

```
    T T H
×   Y A T
  ──────────
    T T H
    A H G
  N T G
  ──────────
  N H B A H
```

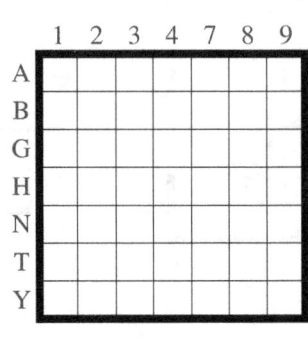

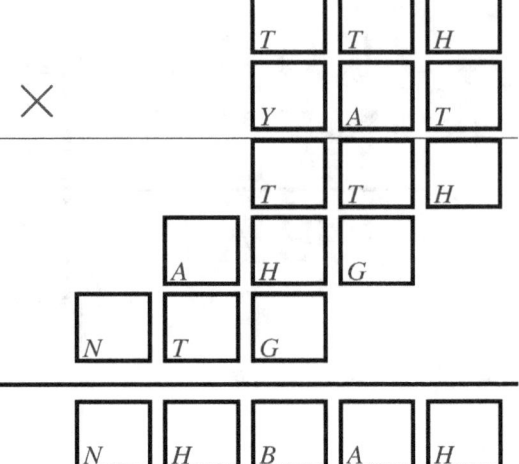

Problem No. 138:

```
    B K B
    E B B
+   T T K
  ──────────
  E B K T
```

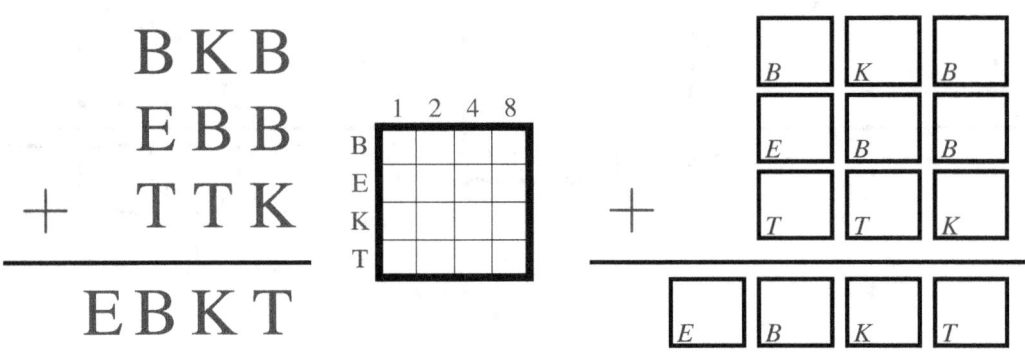

Problem No. 139:

```
  B K H K
  X G G H
+ Y B Y G
──────────
H G N G X
```

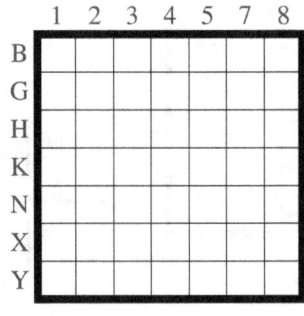

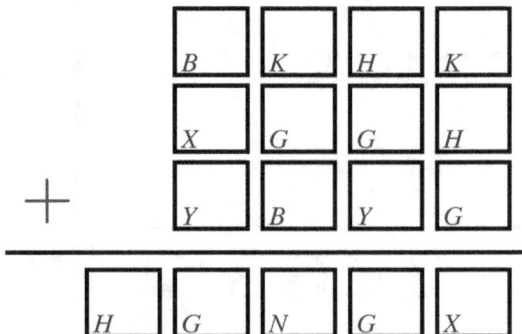

Problem No. 140:

$$
\begin{array}{r}
A\,Y\,E \\
\times \quad A\,Y \\
\hline
T\,T\,B\,T \\
N\,E\,N\,H \\
\hline
K\,K\,B\,N\,T
\end{array}
$$

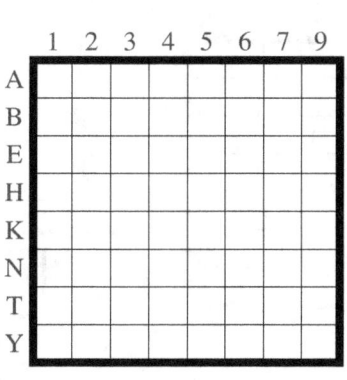
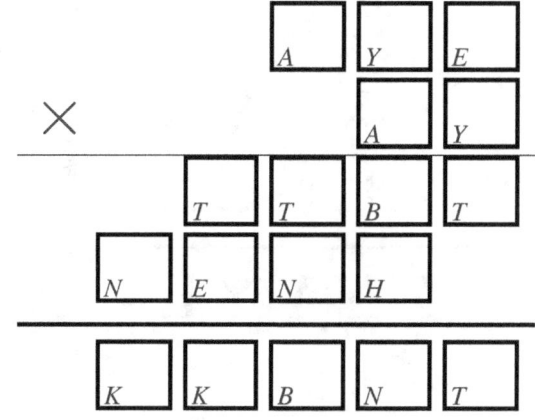

Problem No. 141:

$$
\begin{array}{r}
H\,G\,B\,K \\
B\,B\,X\,K \\
+ \quad X\,E\,A\,X \\
\hline
A\,X\,B\,A\,G
\end{array}
$$

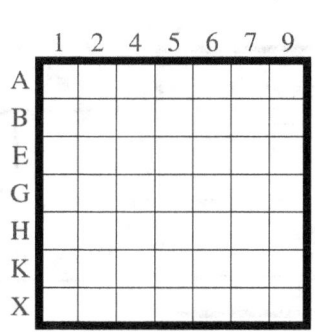
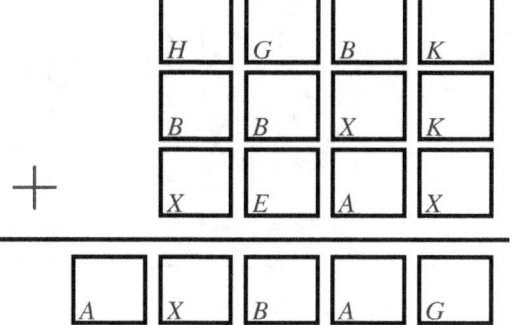

Problem No. 142:

$$
\begin{array}{r}
K\,K \\
\times \quad H\,N \\
\hline
H\,A\,H \\
B\,X\,K \\
\hline
B\,K\,Y\,H
\end{array}
$$

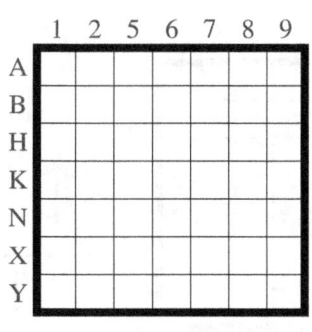
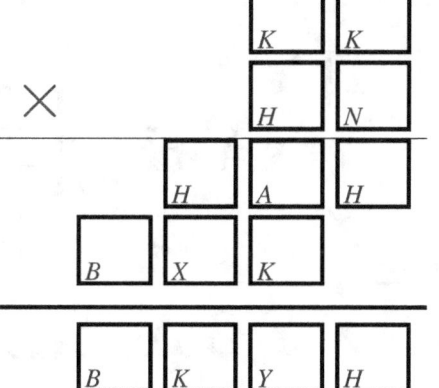

Problem No. 143:

```
  H B N G
  X Y Y X
+ H N G X
─────────
Y N Y Y E
```

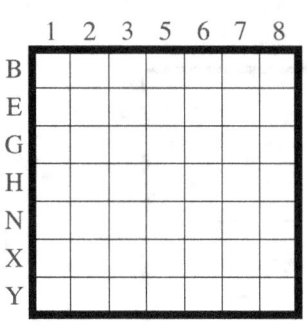
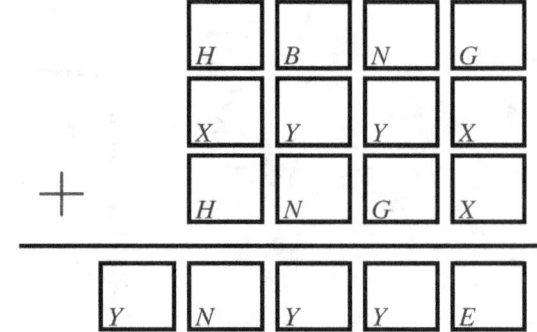

Problem No. 144:

```
    K K G
×     N E
─────────
  X T Y G
  Y Y G
─────────
  E N Y G
```

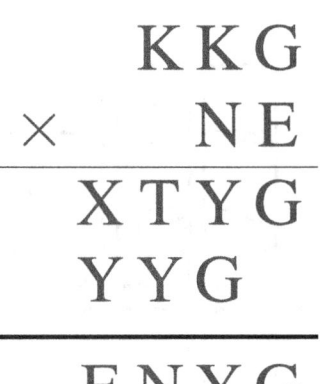
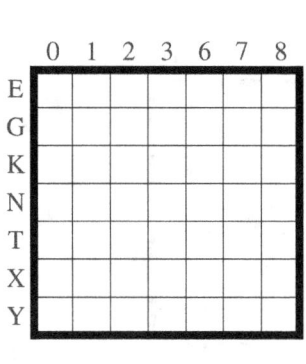
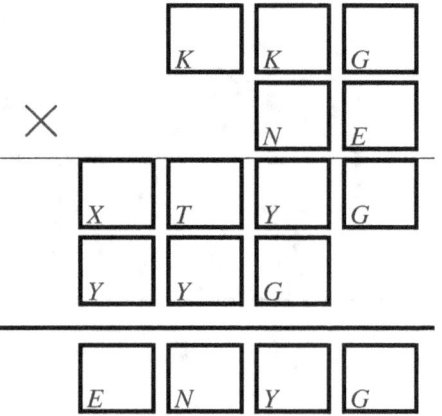

Problem No. 145:

```
  H Y A A
  Y X Y H
+ X T N X
─────────
N X Y B A
```

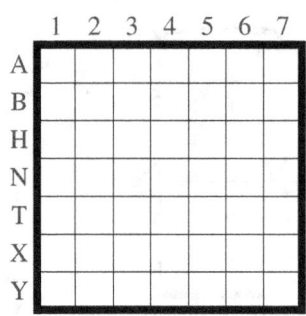
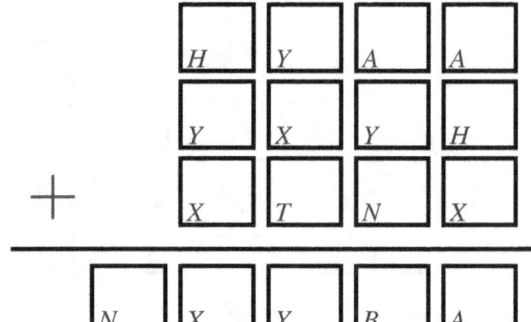

Problem No. 146:

```
      A Y Y
  ×     E E
  ─────────
    K N N H
  K N N H
  ─────────
  K A G E H
```

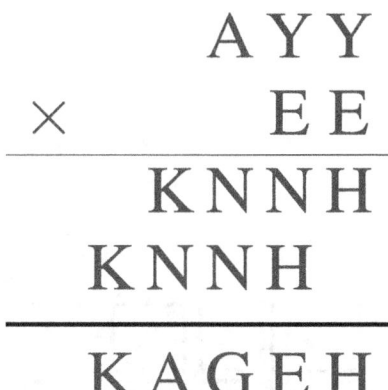
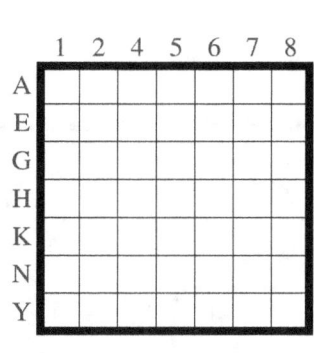
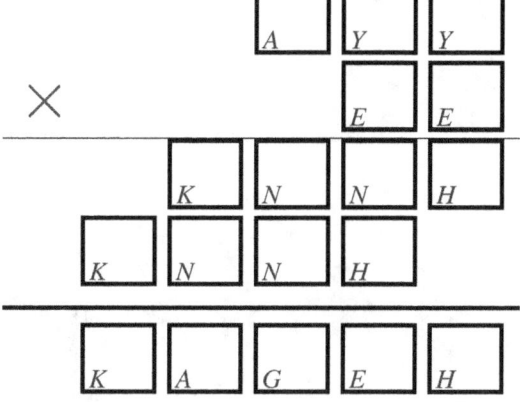

Problem No. 147:

```
      Y H H
  ×     H K
  ─────────
    G K K H
  T X X K
  ─────────
  H T H A H
```

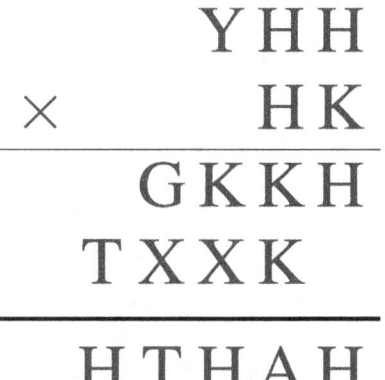
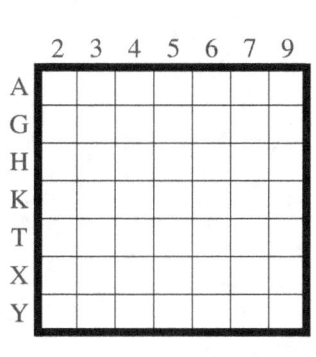
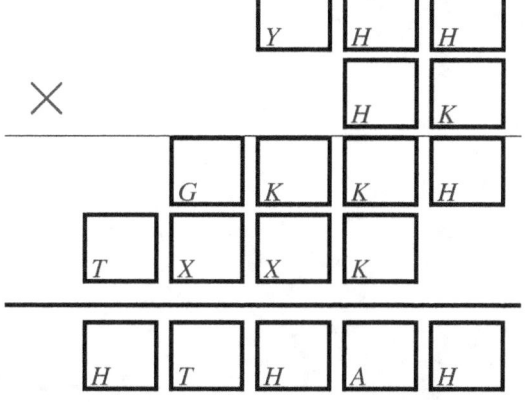

Problem No. 148:

```
      K K G
  ×     G A
  ─────────
    G G N B
  T T K
  ─────────
  N N B A B
```

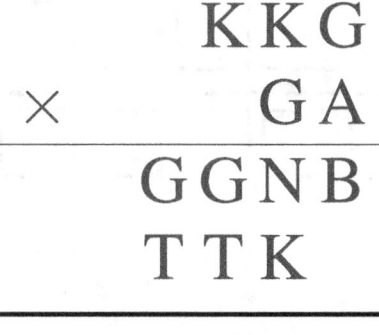
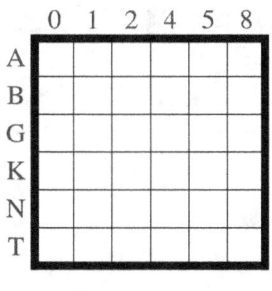
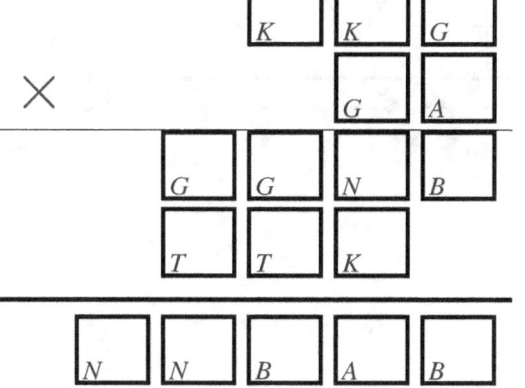

Problem No. 149:

$$
\begin{array}{r}
H\,Y\,H\,G \\
+\quad Y\,K\,T\,N \\
\hline
H\,X\,N\,N\,X
\end{array}
$$

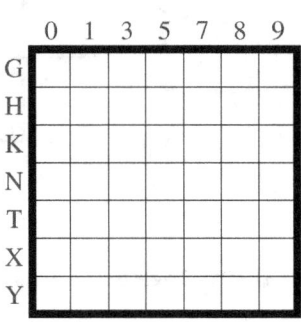

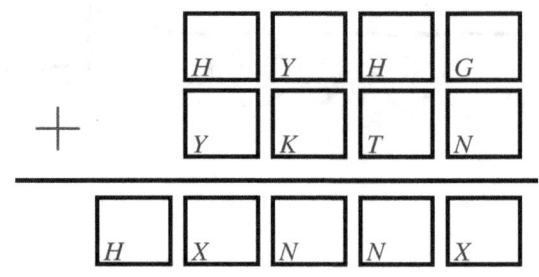

Problem No. 150:

$$
\begin{array}{r}
T\,E\,G\,B \\
N\,K\,N\,G \\
+\quad N\,T\,H\,T \\
\hline
B\,T\,G\,K\,E
\end{array}
$$

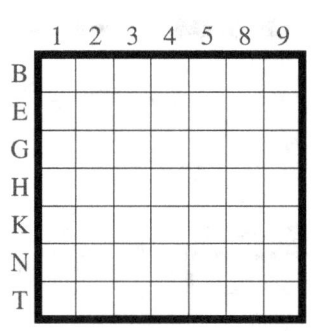

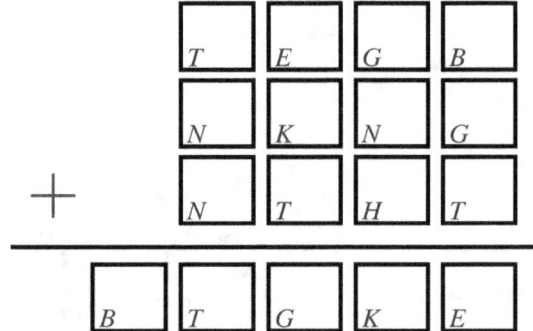

DIFFICULT

Problem No. 151:

```
  X G N A
  H N N H
+ X A N H
─────────
A B N B B
```

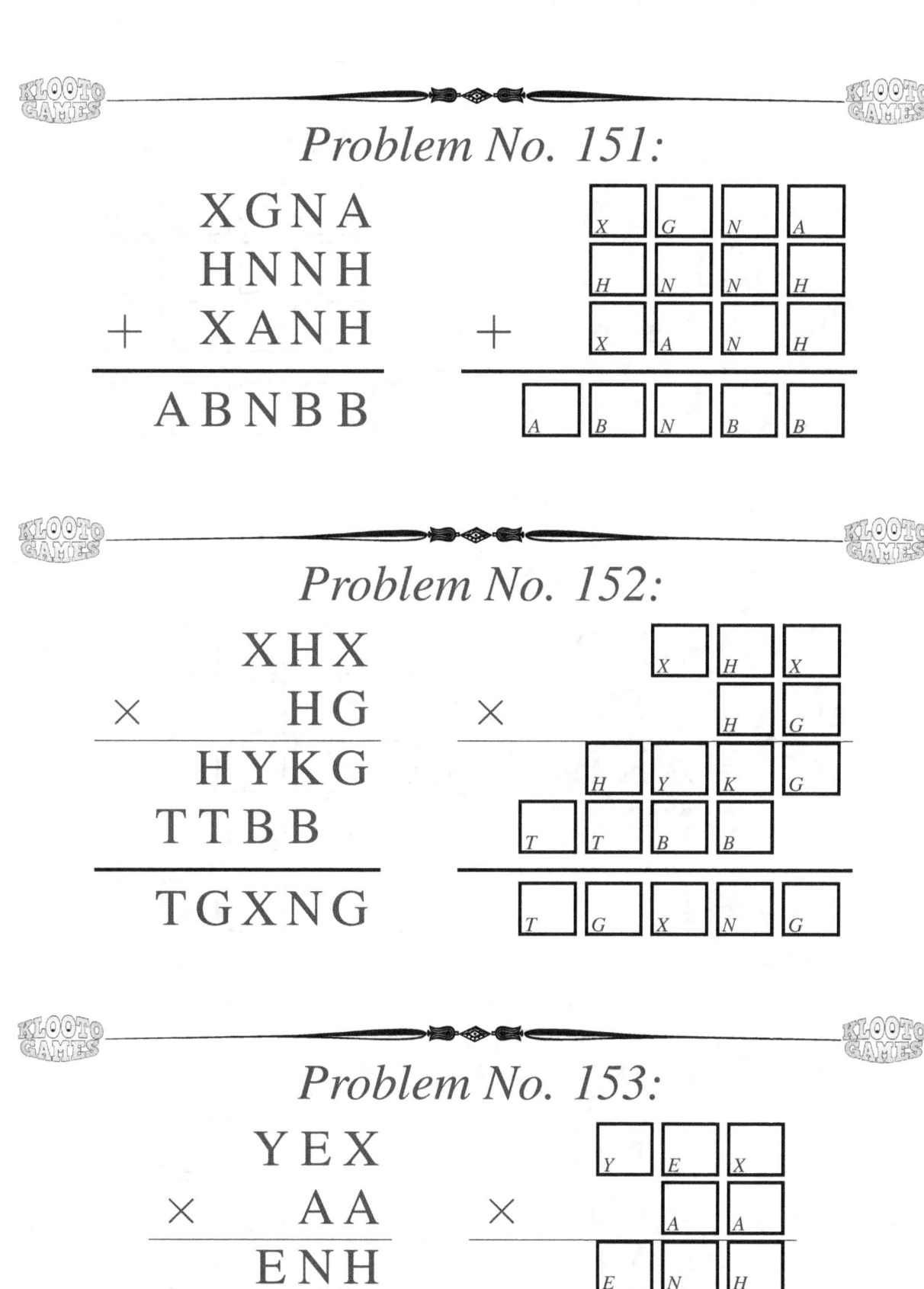

Problem No. 152:

```
    X H X
×     H G
─────────
  H Y K G
  T T B B
─────────
T G X N G
```

Problem No. 153:

```
  Y E X
×   A A
───────
  E N H
  E N H
───────
B Y B H
```

Problem No. 154:

```
  X T Y K
  N N T K
+ N N K K
─────────
K X K X X
```

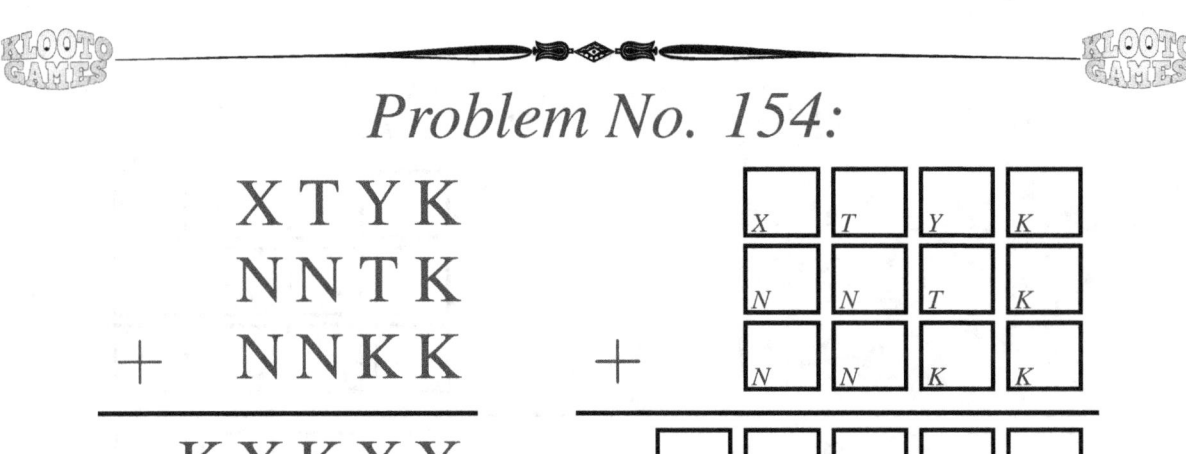

Problem No. 155:

```
    G N B
  ×   E H
  ─────────
  E A K H
  G N B
  ─────────
  H H G H
```

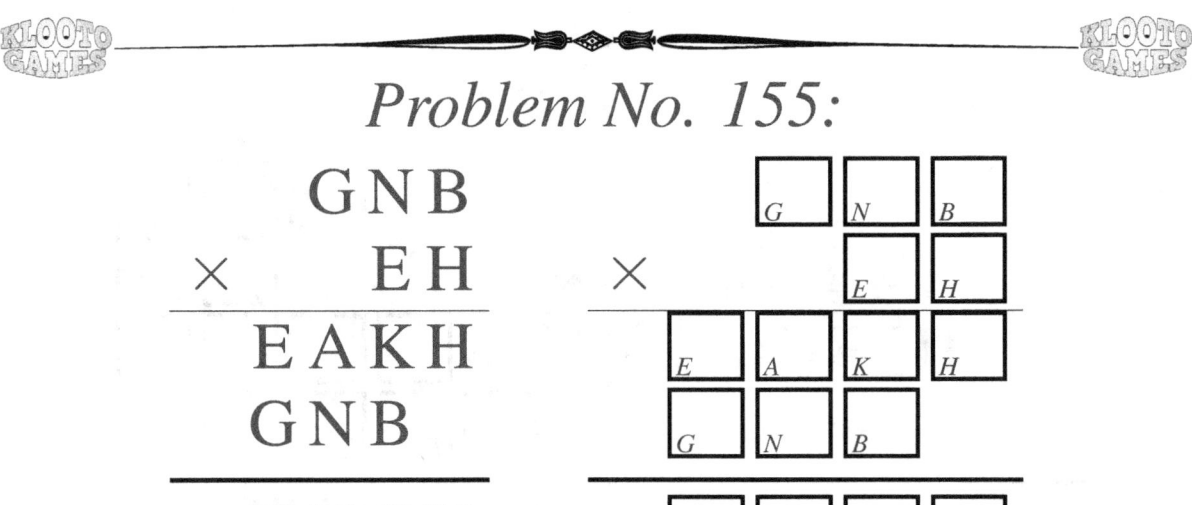

Problem No. 156:

```
      K G
  ×   Y E
  ─────────
    G Y Y
  G B X
  ─────────
  G Y A Y
```

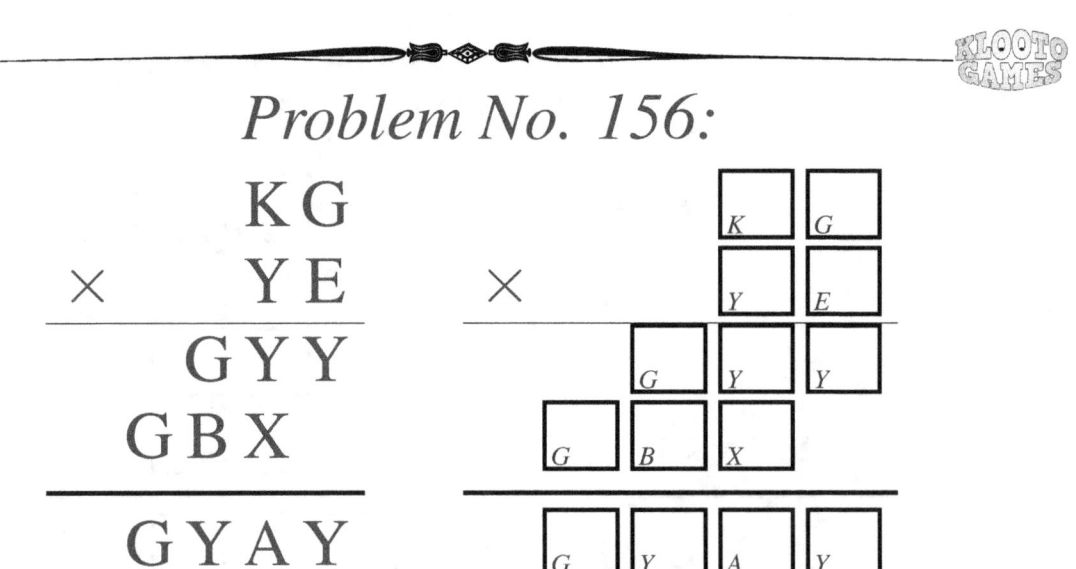

Problem No. 157:

```
      X K Y
  ×     A K
  ─────────
      B E X
    N Y E Y
  ─────────
  N Y B Y X
```

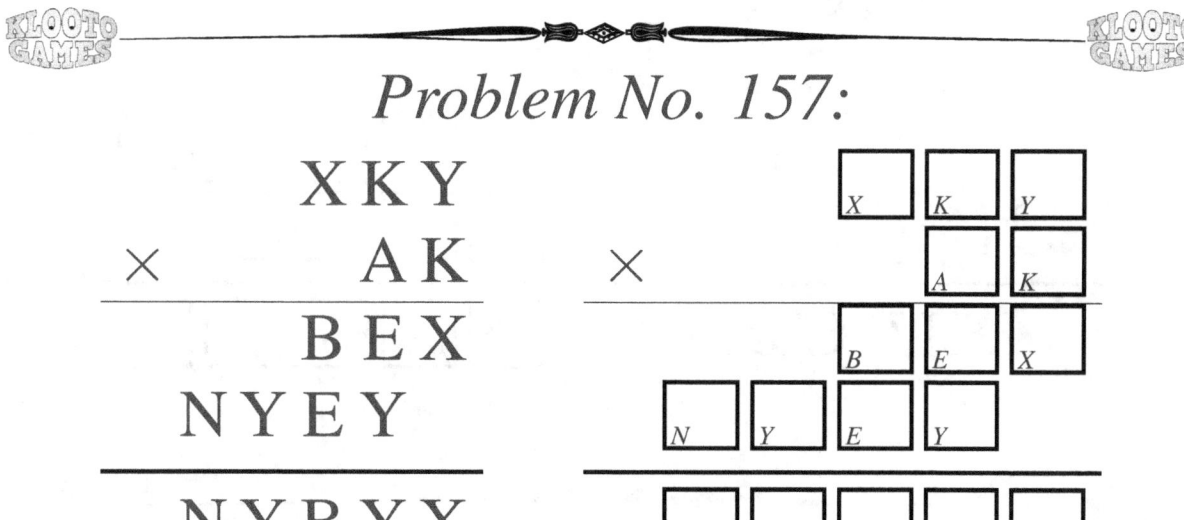

Problem No. 158:

```
      T T B
  ×     H X
  ─────────
    T G A T
  N E B H
  ─────────
  N X B X T
```

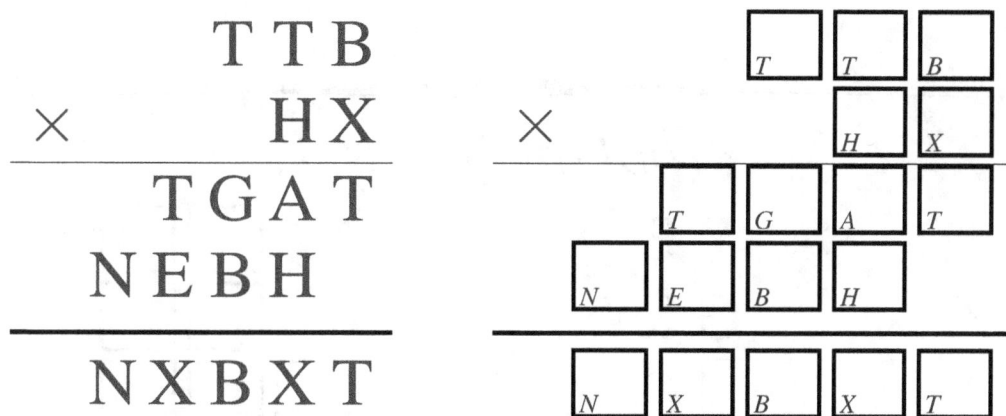

Problem No. 159:

```
    E G B
×     E G
─────────
  K N Y B
N G G B
─────────
E H B G B
```

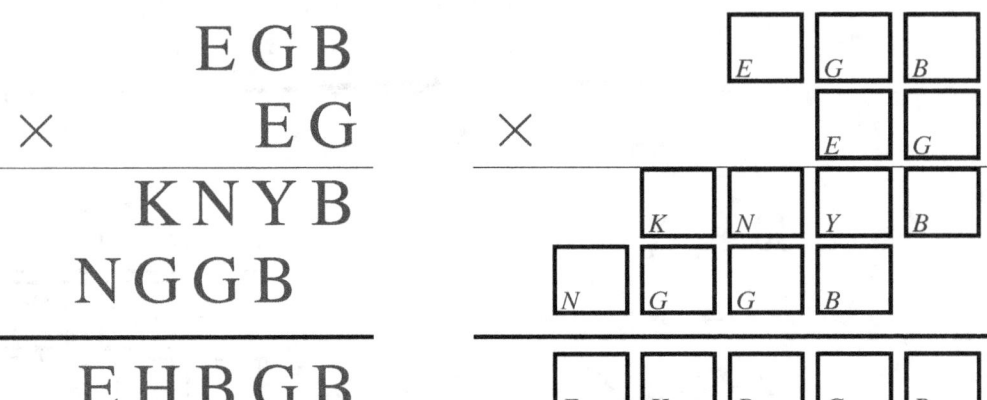

Problem No. 160:

```
  X E X T
  N N T E
+ X G G X
─────────
G G T G N
```

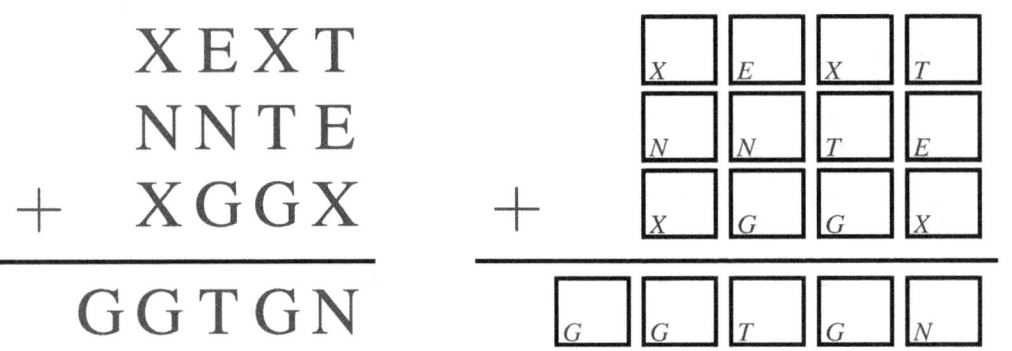

Problem No. 161:

```
    Y B
×   G K
───────
  G N G
N E N
───────
N Y G G
```

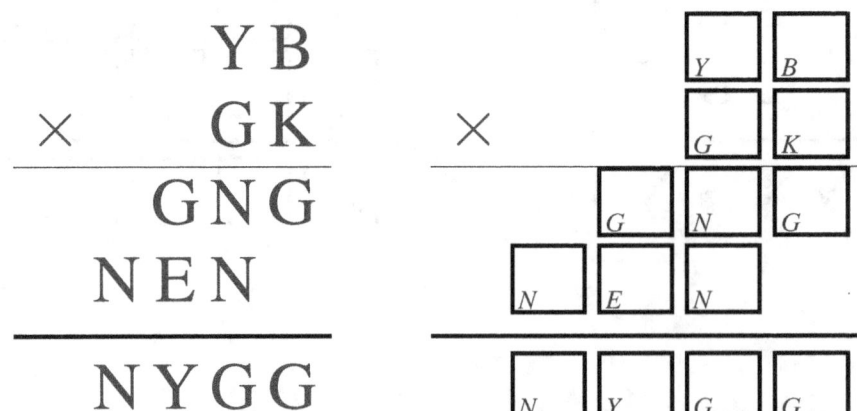

Problem No. 162:

```
  N H K H
  T H N B
+ H N K T
─────────
  K T K K
```

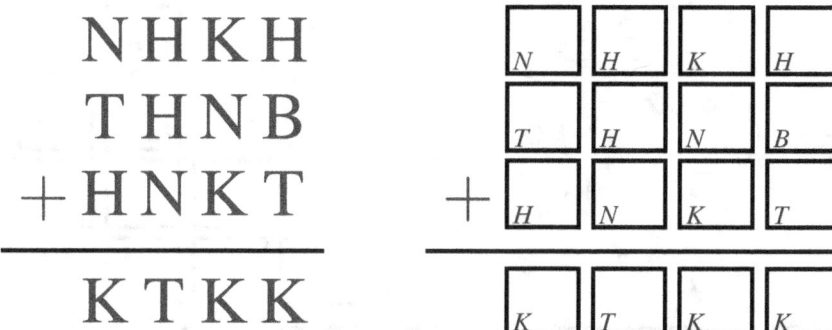

Problem No. 163:

```
      X E B
  ×   X Y K
  ─────────
      B H T K
    B Y X Y
  A K X
  ─────────
  E B A K K
```

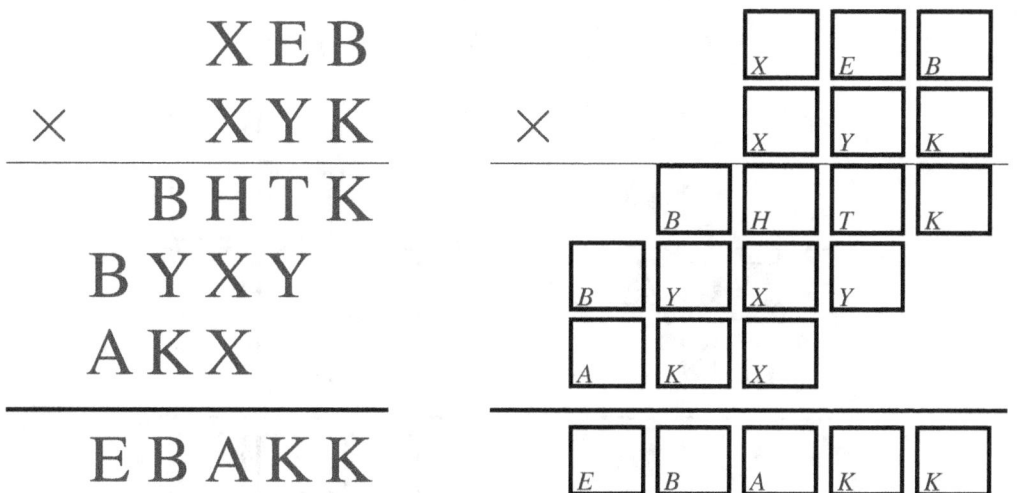

Problem No. 164:

```
  E E T G
  G E Y G
+ K K N N
─────────
  G N G N T
```

Problem No. 165:

```
      E G
  ×   G G
  ─────────
      T Y H
    T Y H
  ─────────
    G K G H
```

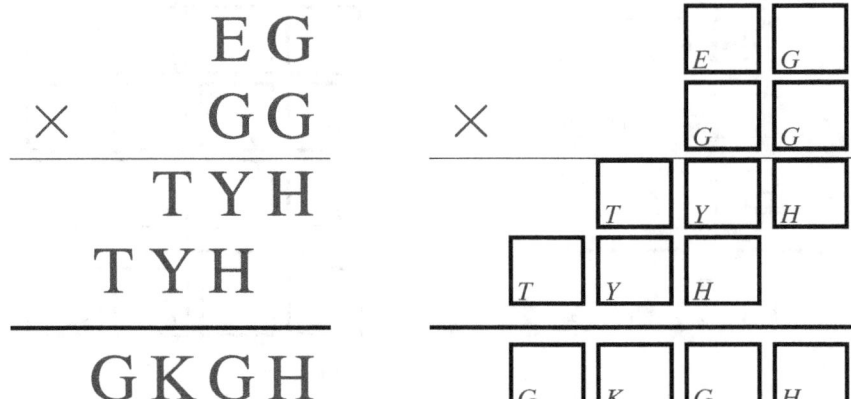

Problem No. 166:

```
      T G
  ×   G H
  ─────────
      E G A
    H N
  ─────────
    A T A
```

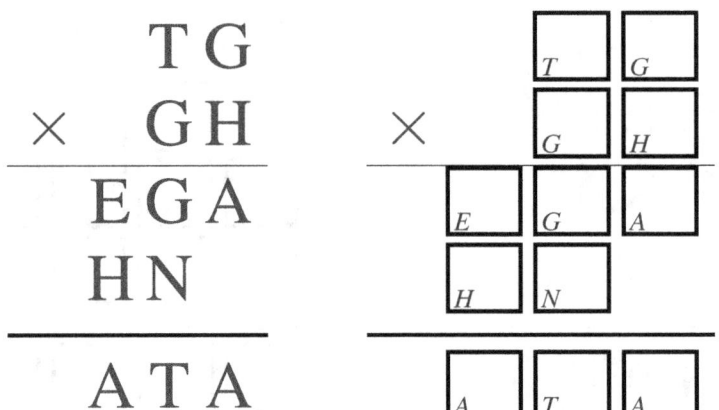

Problem No. 167:

```
      A E N
  ×   X N G
  ─────────
    N A N G
      A E N
  T G X X
  ───────────
  T A N K T G
```

Problem No. 168:

```
  A H X N
  X N N A
+ G A T A
─────────
G H G G T
```

Problem No. 169:

```
  N K B X
  N K T N
+ X N X X
─────────
T B T K Y
```

Problem No. 170:

```
      A B
  ×   G B
  ───────
    G K H
  Y A Y
  ───────
  Y B G H
```

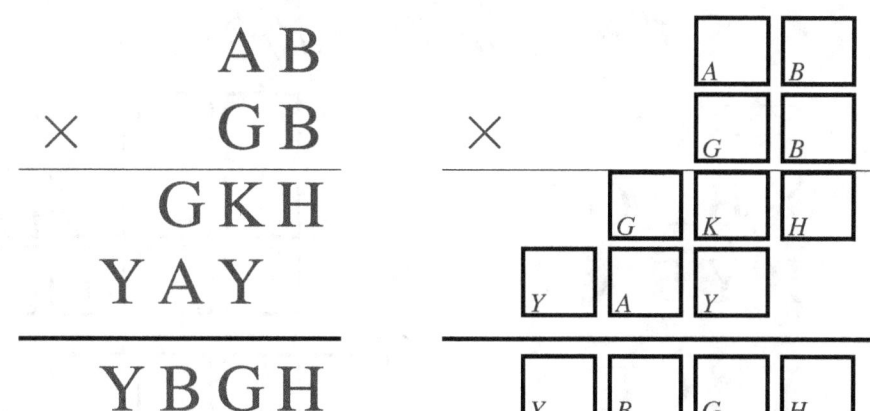

Problem No. 171:

```
      H T
  ×   G T
  ───────
    H N H
  H H Y
  ───────
  H E H H
```

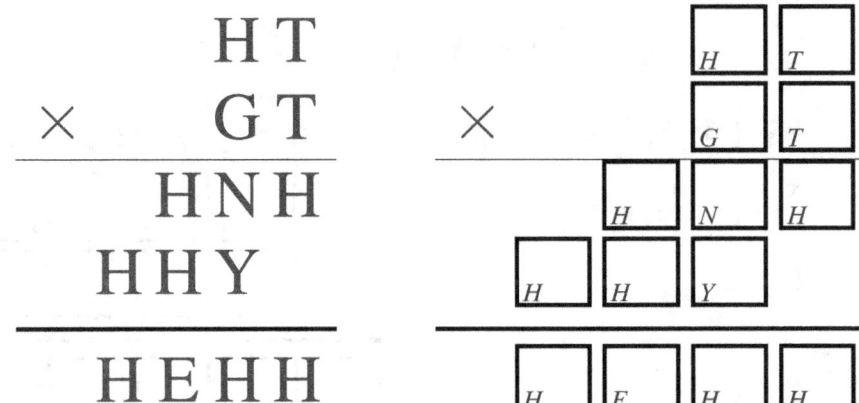

Problem No. 172:

```
    H G A
  ×   E H
  ─────────
    Y N H
  G X K X
  ─────────
  G G H N H
```

Problem No. 173:

```
    X B G Y
    X B B G
  + B Y B X
  ──────────
  G Y X G T
```

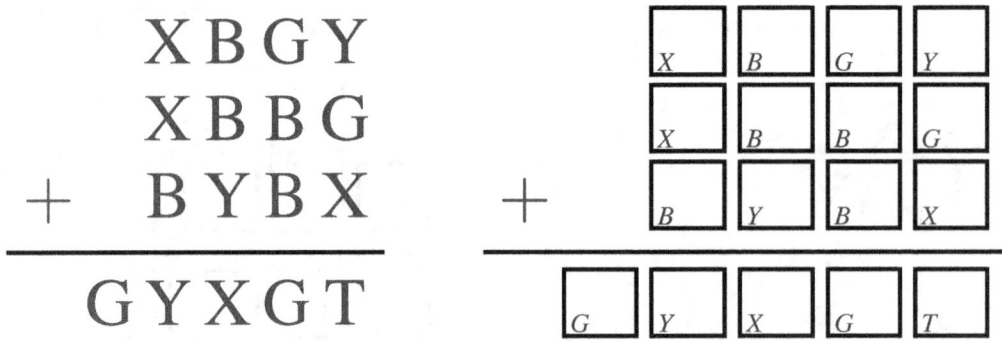

Problem No. 174:

```
  T E X H
  H X Y H
+ E X X K
─────────
K Y X Y Y
```

Problem No. 175:

```
  G T H G
  T H Y Y
+ X X Y T
─────────
B H Y X X
```

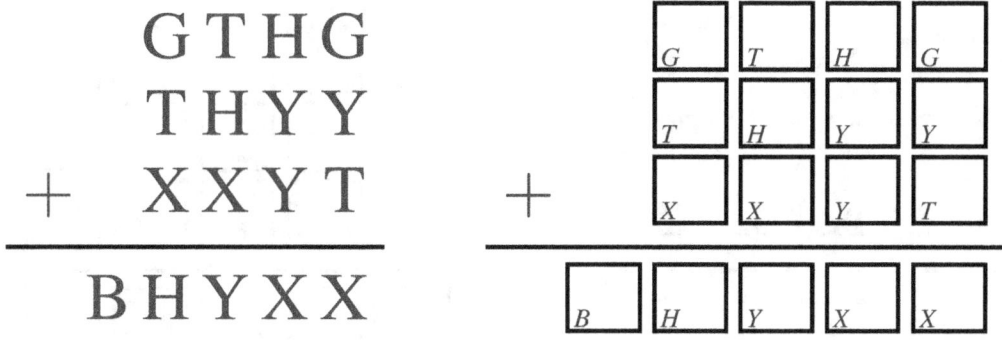

Problem No. 176:

```
  K E H B
  K E X Y
+ E B K K
─────────
X Y E K E
```

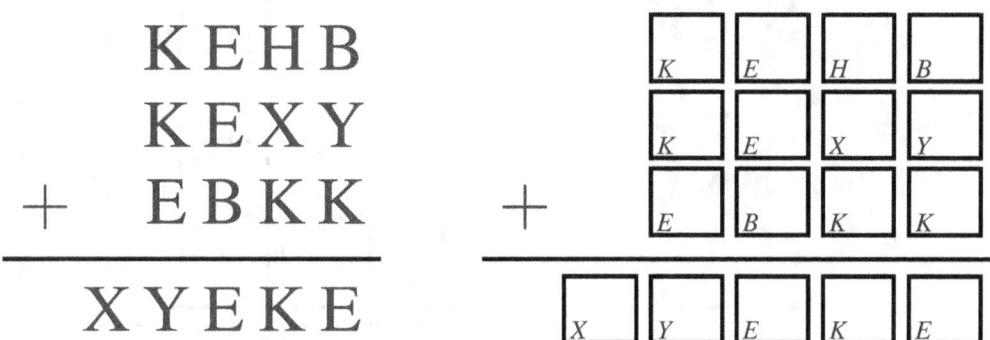

Problem No. 177:

```
        B T
  ×     A H
  ─────────
      N G A
      N A T
  ─────────
      N T H A
```

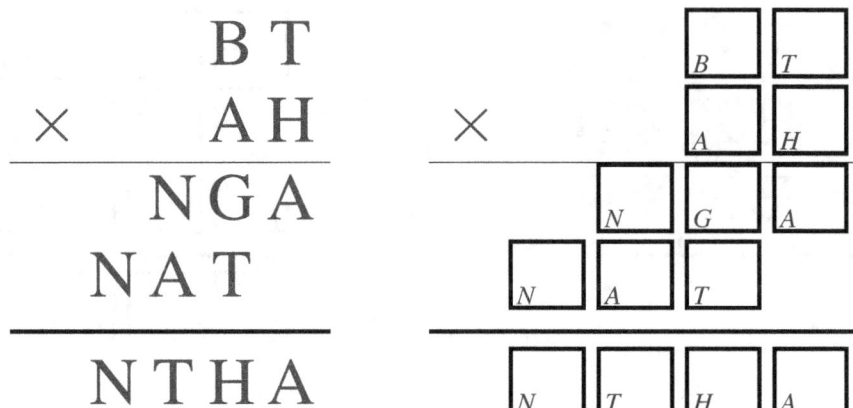

Problem No. 178:

```
      Y Y A Y
      Y A A Y
  +   K K E Y
  ───────────
      E N Y K K
```

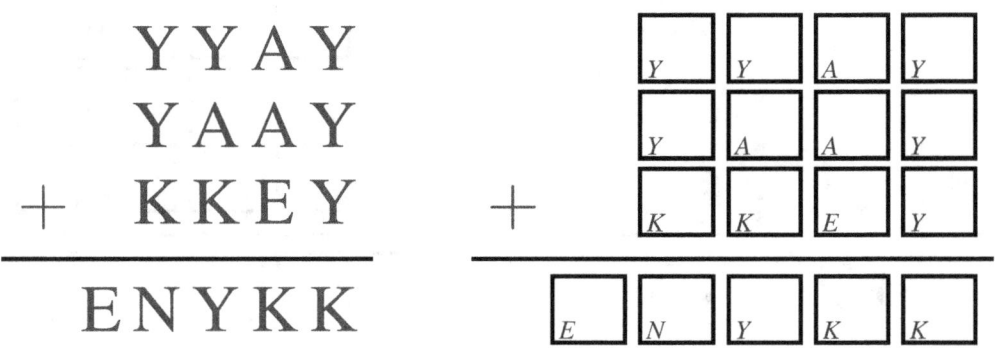

Problem No. 179:

```
      B H K
  ×     Y G
  ─────────
      H A B
    T X H H
  ─────────
    B N Y G B
```

Problem No. 180:

```
        B T E
  ×     E B Y
      ─────────
      G T N G
      G B Y T
    Y T A Y
  ─────────────
    Y N N E N G
```

			B	T	E
			E	B	Y
		G	T	N	G
	G	B	Y	T	
Y	T	A	Y		
Y	N	N	E	N	G

GENIUS LEVEL

Problem No. 181:

```
      B E X
  ×   X E N
  ─────────
    K K K T
    T N K G
  H B G T
  ─────────
  X T T K K T
```

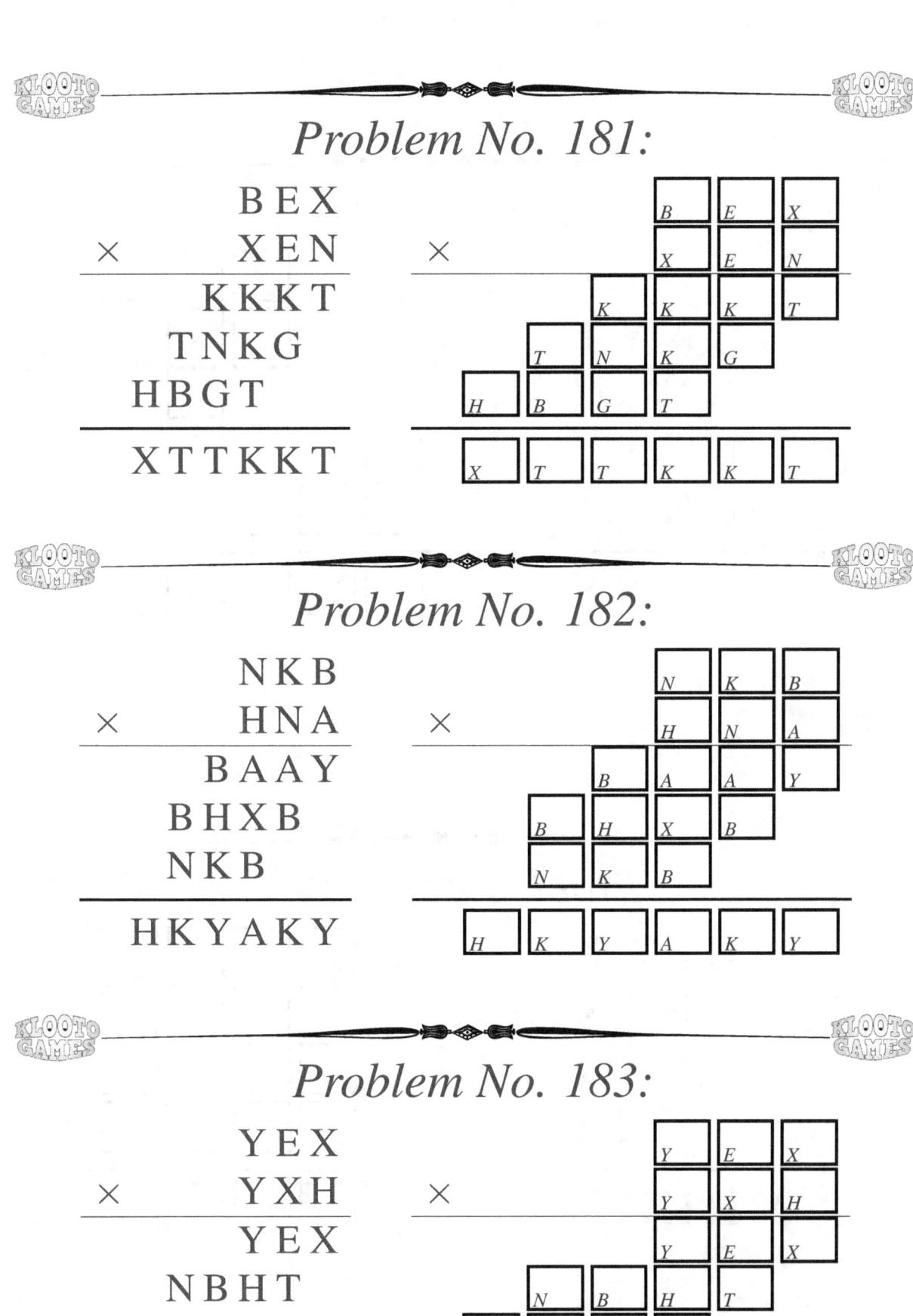

Problem No. 182:

```
      N K B
  ×   H N A
  ─────────
    B A A Y
    B H X B
  N K B
  ─────────
  H K Y A K Y
```

Problem No. 183:

```
      Y E X
  ×   Y X H
  ─────────
    Y E X
    N B H T
  B E B T
  ─────────
  B Y G G H X
```

Problem No. 184:

```
    H Y T
  × N Y Y
  ─────────
  N N G N
  N N G N
  T H K
  ─────────
  A T Y N N
```

		H	Y	T
		N	Y	Y
	N	N	G	N
N	N	G	N	
T	H	K		
A	T	Y	N	N

Problem No. 185:

```
    H X A K
    X H T K
    T H G K
  + A K K A
  ─────────
  H H K K H
```

H	X	A	K
X	H	T	K
T	H	G	K
A	K	K	A

H	H	K	K	H

Problem No. 186:

```
    X X E B
    K X E A
    K A G A
  + N E K X
  ─────────
    K N E N G
```

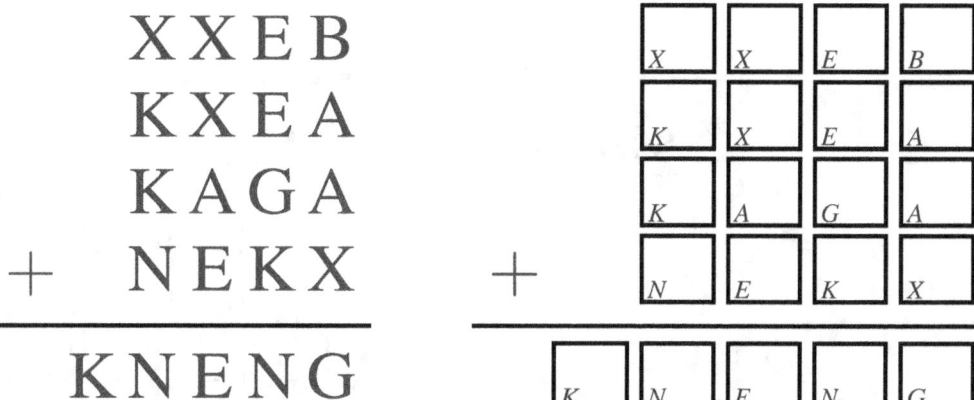

Problem No. 187:

```
    N Y A X
    N N A N
    N N Y X
  + A G G G
  ─────────
    X G Y A K
```

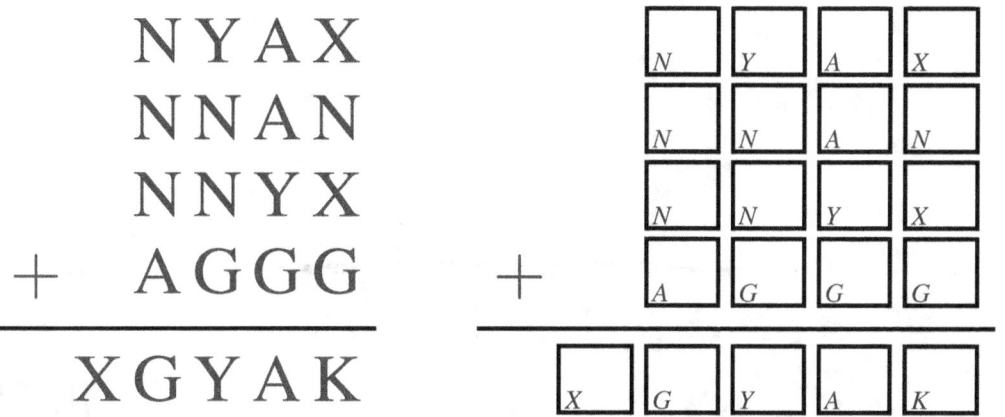

Problem No. 188:

```
    K X K T
    N T N X
    T E E N
  + H X X X
  ─────────
    X K H E E
```

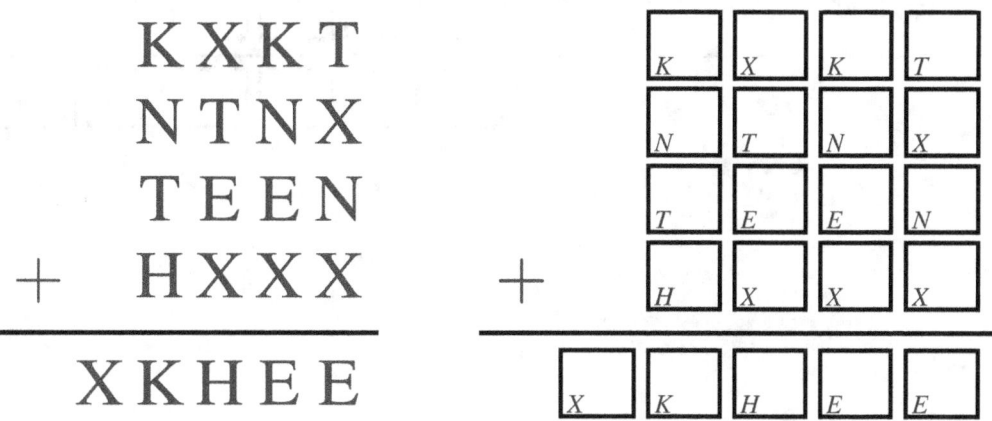

Problem No. 189:

```
        B T B
  ×     T T A
      ─────────
      E A E G
    E N K T
  E N K T
  ─────────────
  E X T T K G
```

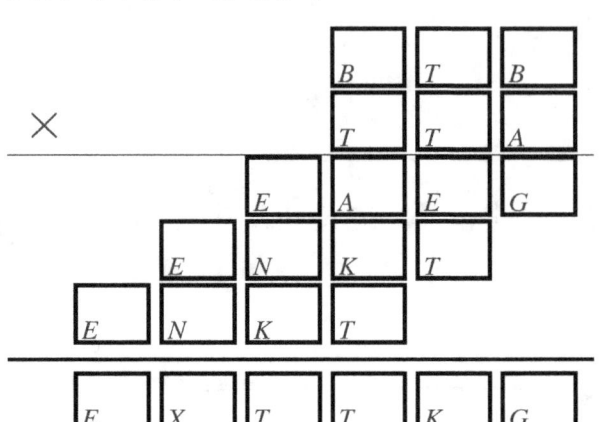

Problem No. 190:

```
        A E X
  ×     E X B
      ─────────
        T N K
      K T E G
    K E N A
  ─────────────
  K X G E G K
```

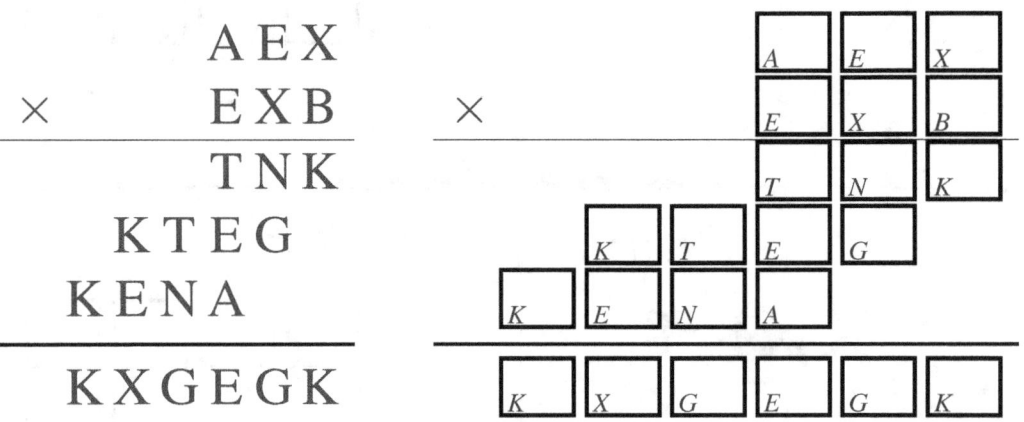

Problem No. 191:

```
  N X A A
  E X X Y
  X Y Y B
+ Y Y Y A
─────────
A N N X N
```

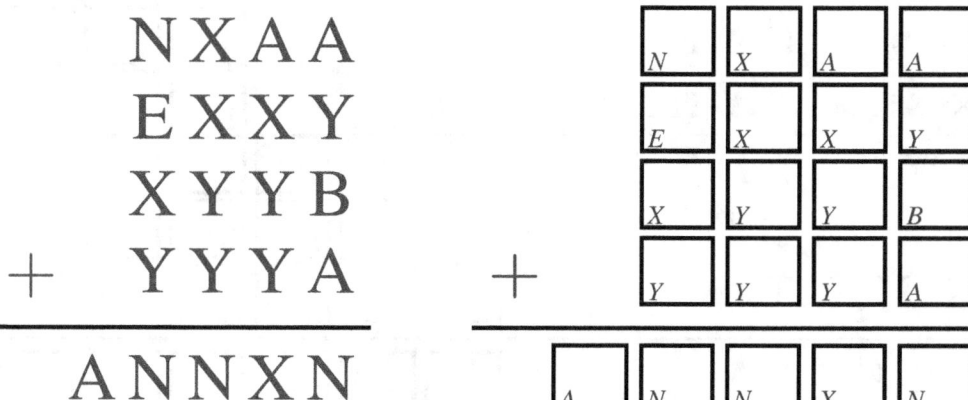

Problem No. 192:

```
  B B T A
  E T E A
  T A G T
+ E X T N
─────────
  X A N T G
```

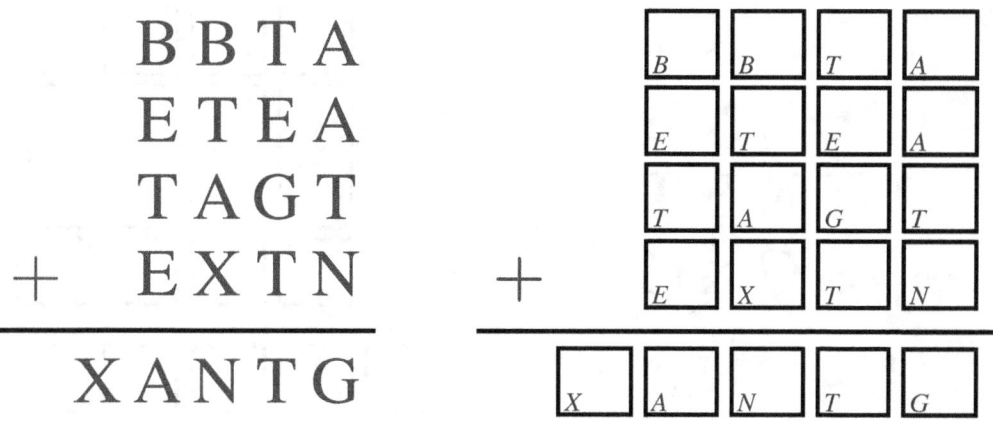

Problem No. 193:

```
      G Y G
  ×   E T N
  ─────────
    Y B Y K
    Y H E X
  Y N G G
  ─────────
  Y X E Y K K
```

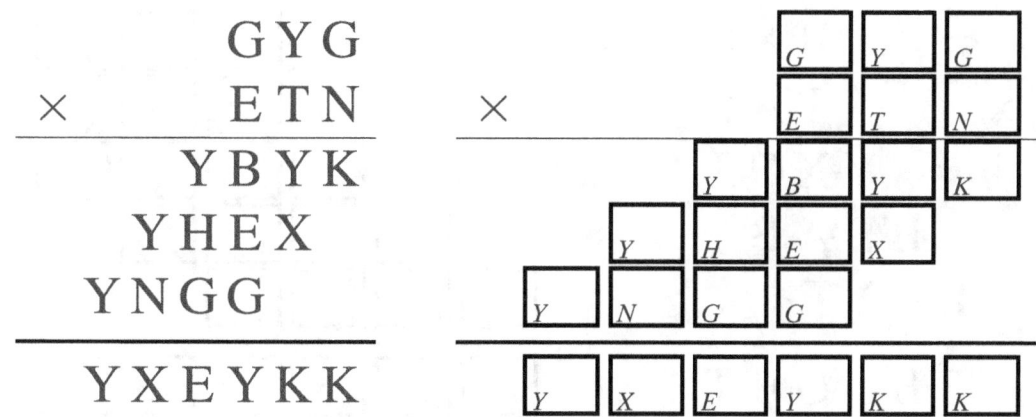

Problem No. 194:

```
        X X K
  ×     Y Y K
  _____
       T E H K
     T X G T
   T X G T
  _____
   T K E Y X K
```

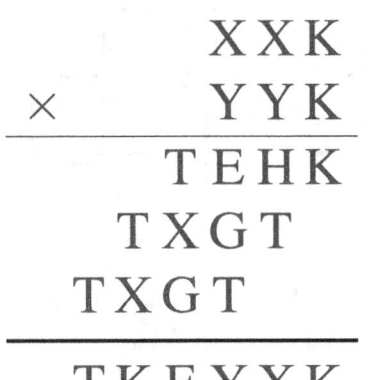

Problem No. 195:

```
      E G N T
      E E N E
      T K E T
  +   G G X G
  _____
      T X T N K
```

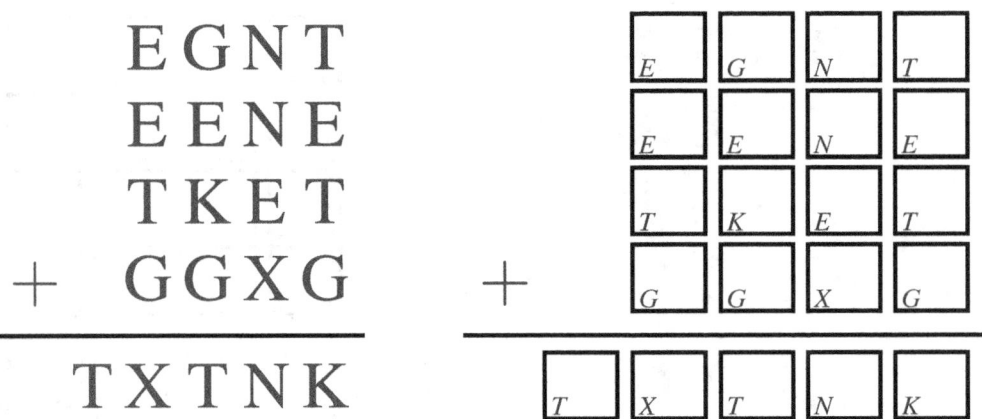

Problem No. 196:

```
        B N N
  ×     T A K
  _____
       Y X G T
     B B A B
   B G Y E
  _____
   B B N G B T
```

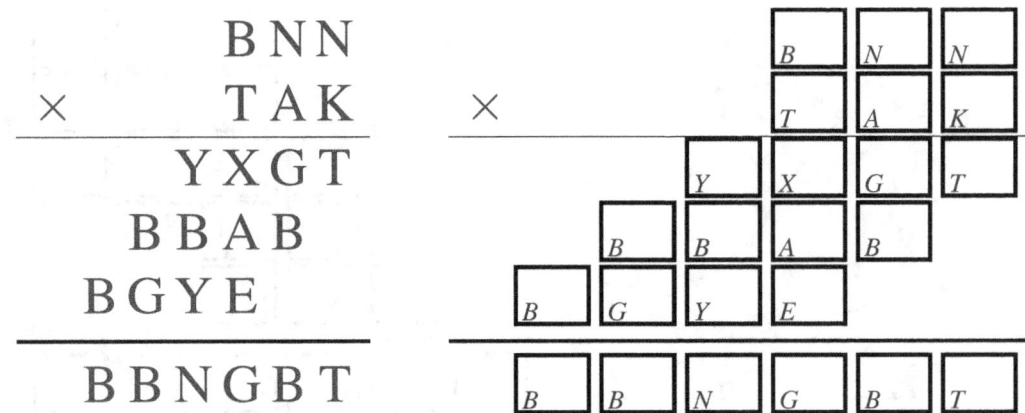

Problem No. 197:

```
      A H K
  ×   A H T
  ─────────
    H E B T
    X E H H
  N H H A
  ─────────
  A B A K H T
```

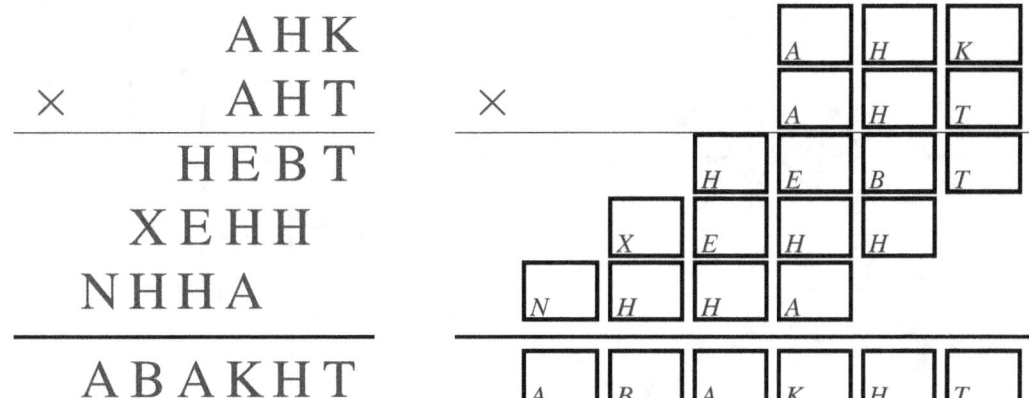

Problem No. 198:

```
      H H K
  ×   T T N
  ─────────
    X N X G
    X X B T
  X X B T
  ─────────
  X H G G G G
```

Problem No. 199:

```
    H N E X
    Y A H N
    Y A X N
  + A E A E
  ─────────
    E N E N E
```

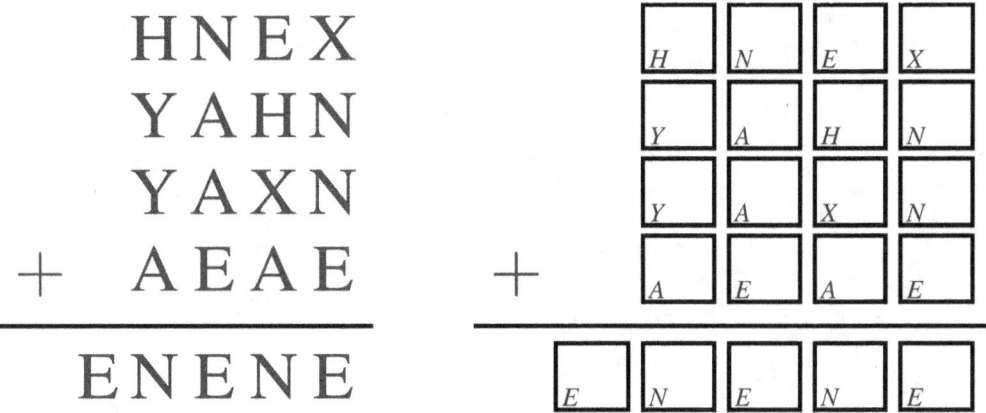

Problem No. 200:

```
        Y K E
  ×     N Y X
  ─────────────
      T K T G
      T Y B Y
      Y K E
  ─────────────
    E G K E G
```

First Order HINTS

These Hints tell you what digits are used for each Problem.
(Still stuck ? Look at the Second Order HINTS that follow!)

(Nos. 1 through 150 already given.)

No. 151: 1 2 4 7 8 9
No. 152: 1 2 3 5 6 7 8 9
No. 153: 1 2 3 4 5 6 7
No. 154: 0 2 4 6 9
No. 155: 1 3 4 5 6 8 9
No. 156: 2 3 4 5 6 8 9
No. 157: 0 1 2 3 4 6 7
No. 158: 0 1 3 4 5 6 8 9
No. 159: 2 4 5 6 7 8 9
No. 160: 0 2 3 4 9
No. 161: 1 2 3 4 5 8
No. 162: 1 2 3 4 8
No. 163: 0 1 2 4 5 6 7 8
No. 164: 1 3 4 5 7 8
No. 165: 0 1 2 4 8 9
No. 166: 1 2 3 6 8 9
No. 167: 1 2 3 4 5 6 7

No. 168: 1 2 3 4 8 9
No. 169: 1 2 3 4 7 8
No. 170: 1 2 3 4 7 9
No. 171: 1 3 4 6 7 9
No. 172: 0 1 2 3 4 5 6 8
No. 173: 1 2 3 5 6
No. 174: 1 3 4 5 7 9
No. 175: 0 1 2 3 5 6
No. 176: 1 2 4 5 7 8
No. 177: 1 2 6 7 8 9
No. 178: 1 2 4 5 8
No. 179: 0 1 2 3 4 5 7 8 9
No. 180: 0 2 3 4 5 6 8
No. 181: 0 1 2 4 5 6 7 9
No. 182: 0 1 3 4 5 7 8
No. 183: 1 3 4 5 6 7 8 9
No. 184: 0 2 3 4 6 7 9
No. 185: 2 3 4 5 7 8

No. 186: 0 1 2 4 6 7 8
No. 187: 1 2 4 7 8 9
No. 188: 1 2 3 4 5 8
No. 189: 1 2 3 4 5 6 7 9
No. 190: 0 1 2 3 6 7 8 9
No. 191: 1 2 3 4 5 9
No. 192: 0 1 2 3 5 6 8
No. 193: 0 1 2 4 5 6 7 8 9
No. 194: 0 1 2 3 5 6 7
No. 195: 0 2 4 5 8 9
No. 196: 0 1 3 4 5 6 7 8 9
No. 197: 0 1 4 5 6 7 8 9
No. 198: 1 2 3 5 6 7 8
No. 199: 2 3 4 6 8 9
No. 200: 0 1 2 3 4 5 7 8

Second Order HINTS
These Hints give you the value for a letter. Good luck!

No. 1: T = 2	No. 52: K = 5	No. 103: T = 4	No. 154: T = 4
No. 2: T = 1	No. 53: E = 6	No. 104: Y = 8	No. 155: G = 3
No. 3: N = 9	No. 54: Y = 8	No. 105: T = 2	No. 156: G = 2
No. 4: K = 4	No. 55: H = 7	No. 106: Y = 4	No. 157: B = 7
No. 5: K = 5	No. 56: T = 5	No. 107: X = 2	No. 158: X = 9
No. 6: K = 3	No. 57: B = 1	No. 108: X = 3	No. 159: H = 4
No. 7: Y = 7	No. 58: N = 2	No. 109: H = 2	No. 160: G = 2
No. 8: T = 1	No. 59: E = 1	No. 110: K = 5	No. 161: Y = 5
No. 9: H = 6	No. 60: X = 7	No. 111: H = 3	No. 162: B = 3
No. 10: E = 0	No. 61: E = 7	No. 112: H = 9	No. 163: T = 8
No. 11: H = 6	No. 62: N = 7	No. 113: H = 2	No. 164: Y = 3
No. 12: H = 4	No. 63: N = 6	No. 114: G = 2	No. 165: G = 2
No. 13: H = 1	No. 64: X = 4	No. 115: B = 4	No. 166: E = 1
No. 14: Y = 4	No. 65: H = 2	No. 116: X = 5	No. 167: T = 2
No. 15: X = 9	No. 66: Y = 3	No. 117: T = 1	No. 168: X = 9
No. 16: X = 5	No. 67: A = 2	No. 118: X = 5	No. 169: B = 1
No. 17: A = 1	No. 68: E = 1	No. 119: A = 6	No. 170: Y = 1
No. 18: Y = 4	No. 69: X = 4	No. 120: N = 9	No. 171: H = 1
No. 19: H = 0	No. 70: G = 1	No. 121: X = 2	No. 172: G = 1
No. 20: Y = 2	No. 71: G = 8	No. 122: X = 2	No. 173: T = 6
No. 21: H = 9	No. 72: H = 0	No. 123: N = 8	No. 174: K = 1
No. 22: T = 1	No. 73: N = 6	No. 124: E = 4	No. 175: T = 2
No. 23: H = 9	No. 74: N = 3	No. 125: B = 8	No. 176: B = 5
No. 24: K = 9	No. 75: G = 8	No. 126: X = 7	No. 177: A = 6
No. 25: X = 5	No. 76: E = 1	No. 127: A = 2	No. 178: K = 4
No. 26: N = 9	No. 77: B = 7	No. 128: E = 5	No. 179: T = 1
No. 27: G = 5	No. 78: E = 9	No. 129: H = 7	No. 180: G = 2
No. 28: X = 2	No. 79: X = 2	No. 130: B = 2	No. 181: X = 2
No. 29: K = 5	No. 80: N = 2	No. 131: K = 8	No. 182: H = 1
No. 30: B = 9	No. 81: X = 4	No. 132: H = 2	No. 183: Y = 9
No. 31: N = 9	No. 82: X = 6	No. 133: B = 3	No. 184: G = 0
No. 32: Y = 7	No. 83: T = 7	No. 134: X = 5	No. 185: K = 5
No. 33: X = 7	No. 84: T = 1	No. 135: H = 7	No. 186: B = 2
No. 34: T = 9	No. 85: K = 6	No. 136: X = 7	No. 187: K = 8
No. 35: G = 1	No. 86: B = 6	No. 137: H = 4	No. 188: K = 5
No. 36: T = 5	No. 87: Y = 5	No. 138: T = 8	No. 189: K = 6
No. 37: T = 1	No. 88: B = 5	No. 139: K = 7	No. 190: E = 6
No. 38: B = 5	No. 89: E = 9	No. 140: H = 6	No. 191: N = 5
No. 39: E = 8	No. 90: G = 1	No. 141: B = 2	No. 192: E = 3
No. 40: E = 8	No. 91: G = 3	No. 142: A = 1	No. 193: G = 4
No. 41: K = 8	No. 92: G = 1	No. 143: Y = 2	No. 194: X = 3
No. 42: Y = 9	No. 93: X = 1	No. 144: E = 8	No. 195: T = 2
No. 43: X = 1	No. 94: T = 5	No. 145: A = 7	No. 196: X = 5
No. 44: N = 9	No. 95: E = 4	No. 146: A = 5	No. 197: T = 6
No. 45: A = 2	No. 96: A = 9	No. 147: H = 4	No. 198: K = 7
No. 46: Y = 5	No. 97: E = 1	No. 148: A = 5	No. 199: E = 3
No. 47: X = 9	No. 98: E = 0	No. 149: T = 5	No. 200: X = 4
No. 48: T = 2	No. 99: B = 8	No. 150: G = 2	
No. 49: Y = 6	No. 100: E = 0	No. 151: H = 8	
No. 50: Y = 4	No. 101: Y = 4	No. 152: H = 3	
No. 51: Y = 5	No. 102: Y = 7	No. 153: A = 4	

No. 1: A1, T2, E6, H7, G8, B9
No. 2: G0, T1, N3, B7, A8
No. 3: K1, X2, G4, B6, A8, N9
No. 4: G2, B3, K4, T6, E7, X9
No. 5: A0, T4, K5, X9
No. 6: X2, K3, Y4, H6, A7, N8
No. 7: A1, K5, B6, Y7
No. 8: T1, X2, A4, G6, N8, E9
No. 9: K1, T2, G3, X4, H6, N8
No. 10: E0, G1, B2, A8
No. 11: T1, Y2, K3, B4, H6, X7, G9
No. 12: X1, E2, N3, H4, B9
No. 13: H1, A3, K5, T8, Y9
No. 14: T1, E2, Y4, X6, A7, N9
No. 15: A1, N2, Y4, G5, B6, X9
No. 16: G0, K1, E4, X5, B6, T8
No. 17: A1, T2, H4, Y5, G7, B8
No. 18: T1, X3, Y4, K6, G8, H9
No. 19: H0, G1, Y5, A7, B8, K9
No. 20: N1, Y2, X3, T4, B7, A9
No. 21: Y1, B2, T3, A5, H9
No. 22: T1, A2, E5, G6, X8
No. 23: B1, T3, X6, K7, H9
No. 24: B1, G3, E5, H7, T8, K9
No. 25: N1, K3, E4, X5, G6, T7, Y8
No. 26: G1, A3, E4, K5, Y7, N9
No. 27: K1, Y2, A3, G5, X8, N9
No. 28: B0, E1, X2, K3, H4, N8
No. 29: G0, N1, B4, K5, A6, Y8
No. 30: N0, Y1, E2, K4, H6, B9
No. 31: G1, K2, B3, T4, N9
No. 32: T0, E1, A3, B4, N6, Y7
No. 33: T1, E2, H3, N5, B6, X7, G9
No. 34: G1, B2, H3, Y6, E7, T9
No. 35: G1, A2, N4, T7, B9
No. 36: X1, Y2, N3, B4, T5, H6
No. 37: T1, N2, B5, A7, Y8, G9
No. 38: A2, H3, B5, G6, Y8
No. 39: N1, Y2, K4, B5, E8
No. 40: N1, Y2, G4, A6, E8
No. 41: B1, X2, H4, Y5, E7, K8
No. 42: A2, X3, B5, T6, H7, E8, Y9
No. 43: X1, T2, Y5, N6, B7, A9
No. 44: B1, Y2, A3, G5, K7, N9
No. 45: Y1, A2, T3, B5, G7
No. 46: A1, H2, X3, Y5, K7
No. 47: K1, N2, B4, G7, X9
No. 48: X1, T2, K6, B7
No. 49: G0, H1, A2, B4, Y6, N8, K9
No. 50: B1, G2, K3, Y4, E5, N8, H9
No. 51: N1, X2, T4, Y5, K7, B8, H9
No. 52: G1, T2, E4, K5, N6, X8
No. 53: N1, Y2, G5, E6, B7
No. 54: H1, N2, K3, X4, A5, Y8, T9
No. 55: K1, X2, E3, A6, H7, G8
No. 56: Y1, X2, E4, T5, B7
No. 57: B1, G2, E3, T5, K6, Y7, A9
No. 58: E1, N2, T3, B4, Y5, H6, K7

No. 59: E1, B2, K5, A8, H9
No. 60: G1, T3, B4, Y5, H6, X7, E9
No. 61: Y1, T3, H4, N5, B6, E7, K8, X9
No. 62: Y1, H2, T4, X6, N7, E8
No. 63: E1, Y2, K4, N6, T7, H8
No. 64: T1, H2, X4, G8
No. 65: G1, H2, N4, K7, B9
No. 66: T1, N2, Y3, G6, B9
No. 67: B0, H1, A2, Y5, N6, X7, G8, T9
No. 68: E1, T2, K3, H4, N5, G6, B8
No. 69: K1, H2, G3, X4, Y5, B7, E8, N9
No. 70: G1, X2, H6, T7, A8
No. 71: Y0, X1, B2, E5, H7, G8, A9
No. 72: H0, K1, A3, E5, X6, G8, N9
No. 73: E0, H1, X2, T4, K5, N6, Y8
No. 74: E0, H1, N3, T4, G6, X7, K9
No. 75: X1, A2, H4, K6, G8
No. 76: E1, H2, K7, Y8, A9
No. 77: H1, T2, K3, G5, B7, A8
No. 78: H1, B2, X4, K6, Y7, T8, E9
No. 79: Y1, X2, E3, G5, B7, H9
No. 80: T1, N2, X3, E7, G9
No. 81: E1, G2, X4, K5, Y6, T7, H8, A9
No. 82: N1, Y2, H3, A5, X6, E7, B8
No. 83: K1, E3, A5, G6, T7, N8, H9
No. 84: T1, G2, A3, B5, Y6, H7, K8, N9
No. 85: E1, H2, G3, N4, Y5, K6, A7
No. 86: Y1, A4, B6, E9
No. 87: H1, T2, Y5, G7, X9
No. 88: N0, X1, K4, B5, G7, Y8, T9
No. 89: X0, Y1, H3, N4, A5, T6, E9
No. 90: B0, G1, X2, K3, E6, T7, H8, N9
No. 91: K1, B2, G3, A4, N5, H6, X8
No. 92: G1, K2, Y5, B6, N7, A8, E9
No. 93: X1, A2, Y3, G6, B8
No. 94: H1, K2, T5, B9
No. 95: A1, K2, X3, E4, T6, B8, N9
No. 96: Y1, K2, N3, B4, H5, G7, E8, A9
No. 97: E1, X2, N3, T5, K6, Y7, G8
No. 98: E0, A1, X2, Y3, T5, K6, B7, G8
No. 99: E1, A2, K4, G5, N6, X7, B8
No. 100: E0, X1, G2, B3, Y6, K7, T9
No. 101: X0, G2, Y4, B5, K6, E9
No. 102: B1, K3, N4, X5, Y7
No. 103: K1, N2, G3, T4, Y9
No. 104: X1, G2, B3, T6, Y8
No. 105: A1, T2, B3, Y4, E6, X7, H8, N9
No. 106: A1, K2, B3, Y4, E6, X7, G9
No. 107: K1, X2, T3, H6, N7, B8
No. 108: G1, T2, X3, Y4, K5, B6, A8
No. 109: H2, B3, K4, T5, X6, G7, A8, E9
No. 110: E1, H2, X3, B4, K5, A8, G9
No. 111: K1, G2, H3, T7, X9
No. 112: N1, Y2, K3, X4, E5, A6, H9
No. 113: G1, H2, B3, X5, N6, K7, A8
No. 114: T1, G2, E3, K6, H7, X8, A9
No. 115: Y1, T2, H3, B4, X5, N6, A7, E8
No. 116: H0, T1, G2, N3, K4, X5, A8, Y9

No. 117: E0, T1, B3, Y4, N6, G7, X8, A9
No. 118: H1, A2, B3, X5, T7
No. 119: N1, H2, B3, K4, X5, A6, T7, G8, E9
No. 120: T1, E2, B3, G4, Y5, A6, N9
No. 121: N1, X2, A5, B7, G8, T9
No. 122: A0, X2, Y6, H7, B8, T9
No. 123: T1, E2, N8, A9
No. 124: K0, X1, T3, E4, Y6, A7, G8
No. 125: N2, T3, K4, G5, E6, B8, A9
No. 126: Y1, G2, A3, K5, B6, X7, H8
No. 127: T1, A2, N4, Y8, E9
No. 128: T1, B2, E5, K9
No. 129: Y1, N2, A3, B4, T6, H7, X9
No. 130: E0, T1, B2, G4, X6, H8, A9
No. 131: A1, H2, N3, B4, T7, K8, Y9
No. 132: A1, H2, B4, E5, K7, X8, G9
No. 133: T0, A1, B3, K4, N6, E7, G9
No. 134: A1, Y4, X5, B6, G7, K8, N9
No. 135: B1, G2, A4, E6, H7, K8
No. 136: G2, H3, B4, K6, X7, N8, T9
No. 137: T1, G2, A3, H4, B7, Y8, N9
No. 138: E1, B2, K4, T8
No. 139: H1, X2, B3, G4, N5, K7, Y8
No. 140: N1, K2, T3, A4, B5, H6, Y7, E9
No. 141: A1, B2, G4, E5, X6, H7, K9
No. 142: A1, X2, B5, H6, N7, K8, Y9
No. 143: E1, Y2, N3, G5, B6, H7, X8
No. 144: G0, X1, K2, N3, Y6, T7, E8
No. 145: N1, B2, Y3, X4, T5, H6, A7
No. 146: N1, G2, K4, A5, H6, E7, Y8
No. 147: A2, T3, H4, G5, K6, X7, Y9
No. 148: B0, N1, G2, K4, A5, T8
No. 149: X0, H1, G3, T5, N7, Y8, K9
No. 150: B1, G2, H3, N4, T5, E8, K9
No. 151: A1, N2, X4, B7, H8, G9
No. 152: B1, T2, H3, G5, Y6, X7, K8, N9
No. 153: Y1, H2, X3, A4, N5, E6, B7
No. 154: Y0, K2, T4, X6, N9
No. 155: E1, G3, K4, H5, N6, A8, B9
No. 156: G2, K3, A4, B5, X6, Y8, E9
No. 157: E0, N1, X2, K3, Y4, A6, B7
No. 158: G0, A1, N3, T4, E5, B6, H8, X9
No. 159: Y2, H4, B5, K6, G7, N8, E9

No. 160: T0, G2, N3, E4, X9
No. 161: E1, N2, B3, G4, Y5, K8
No. 162: T1, N2, B3, H4, K8
No. 163: H0, B1, X2, K4, A5, Y6, E7, T8
No. 164: G1, Y3, K4, N5, T7, E8
No. 165: K0, T1, G2, H4, Y8, E9
No. 166: E1, T2, G3, H6, A8, N9
No. 167: N1, T2, G3, A4, X5, K6, E7
No. 168: G1, A2, H3, N4, T8, X9
No. 169: B1, T2, X3, Y4, K7, N8
No. 170: Y1, K2, G3, A4, B7, H9
No. 171: H1, E3, Y4, G6, N7, T9
No. 172: X0, G1, H2, N3, Y4, E5, A6, K8
No. 173: G1, Y2, X3, B5, T6
No. 174: K1, H3, E4, X5, Y7, T9
No. 175: H0, B1, T2, X3, G5, Y6
No. 176: Y1, X2, E4, B5, H7, K8
No. 177: N1, B2, A6, H7, T8, G9
No. 178: N1, E2, K4, A5, Y8
No. 179: N0, T1, B2, G3, K4, A5, Y7, H8, X9
No. 180: T0, G2, N3, Y4, B5, A6, E8
No. 181: G0, H1, X2, T4, E5, K6, N7, B9
No. 182: Y0, H1, K3, X4, B5, N7, A8
No. 183: H1, N3, X4, E5, T6, G7, B8, Y9
No. 184: G0, N2, H3, K4, Y6, T7, A9
No. 185: H2, G3, X4, K5, A7, T8
No. 186: E0, K1, B2, G4, X6, N7, A8
No. 187: X1, N2, G4, A7, K8, Y9
No. 188: X1, T2, H3, N4, K5, E8
No. 189: E1, G2, B3, A4, T5, K6, N7, X9
No. 190: N0, K1, A2, B3, E6, X7, T8, G9
No. 191: A1, E2, X3, Y4, N5, B9
No. 192: A0, G1, X2, E3, T5, N6, B8
No. 193: K0, B1, Y2, G4, N5, E6, T7, X8, H9
No. 194: E0, H1, T2, X3, G5, K6, Y7
No. 195: N0, T2, X4, E5, K8, G9
No. 196: G0, Y1, B3, K4, X5, E6, N7, T8, A9
No. 197: B0, K1, X4, H5, T6, E7, N8, A9
No. 198: X1, H2, B3, T5, G6, K7, N8
No. 199: N2, E3, H4, X6, A8, Y9
No. 200: K0, N1, T2, B3, X4, Y5, E7, G8

Time To Graduate to CODEWORDS!
YOUR BRAIN WILL GROW

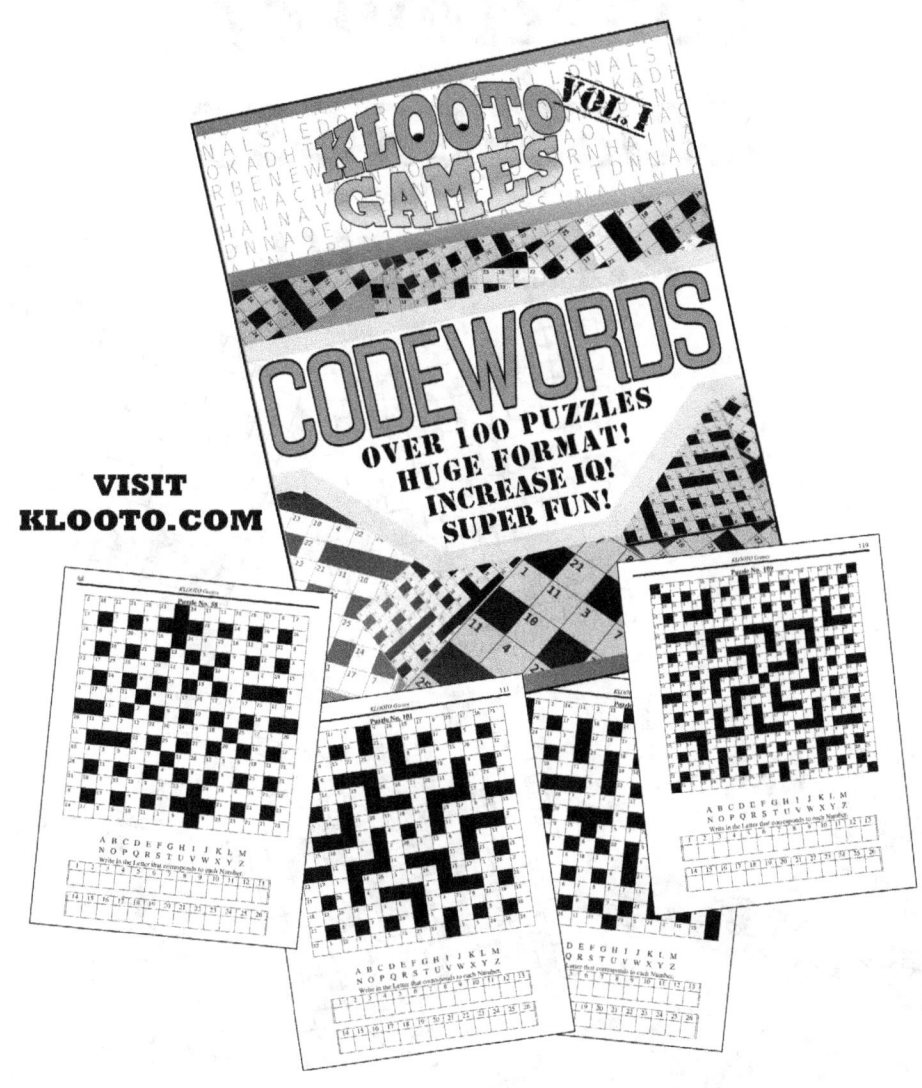

Like Cryptograms?
You'll <u>love</u> KLOOTO's NEW
Collection of
CRISS-CROSS
Puzzles!!!

Buy It at AMAZON!!!
This Ain't Your Grandma's
Criss-Cross....
It Will Fry Your Brain!!!

www.ingramcontent.com/pod-product-compliance
Lightning Source LLC
Chambersburg PA
CBHW080827180526
45168CB00006B/2595